576.6
Esl2v

46470

THE JOHN J. WRIGHT LIBRARY
LA ROCHE COLLEGE
9000 Babcock Boulevard
Pittsburgh, PA 15237

VIRUSES IN PLANT HOSTS

VIRUSES
IN PLANT HOSTS

FORM, DISTRIBUTION, AND

PATHOLOGIC EFFECTS

KATHERINE ESAU

THE 1968 JOHN CHARLES WALKER LECTURES

576.6
Es12v

THE UNIVERSITY OF WISCONSIN PRESS
Madison, Milwaukee, and London, 1968

Published by
The University of Wisconsin Press
Box 1379, Madison, Wisconsin 53701
The University of Wisconsin Press, Ltd.
27–29 Whitfield Street, London, W.1

Copyright © 1968 by
The Regents of the University of Wisconsin
All rights reserved

Manufactured in the United States of America by
Kingsport Press, Inc., Kingsport, Tennessee
Illustration section printed by
Meriden Gravure Co., Meriden, Connecticut
Standard Book Number 299–05110–2
Library of Congress Catalog Card Number 68–9831

46470

The John J. Wright Library
LA ROCHE COLLEGE
9000 Babcock Boulevard
Pittsburgh, Pa. 15237

FOREWORD

A FEW YEARS AGO, friends of John Charles Walker in Racine, Wisconsin, desired to create a lasting memorial to the achievements of one of their most illustrious sons. After considering many kinds of programs, the group agreed upon an endowed lectureship in Walker's name. Under the leadership of the Racine Chamber of Commerce, money was collected from friends and businessmen and given to the Wisconsin Alumni Research Foundation to invest. The earnings are made available to the Department of Plant Pathology for the purpose of bringing eminent scientists to the campus to lecture.

Rather than have a group of unrelated lectures, the Department voted to schedule, as often as funds would permit, J. C. Walker Conferences in Plant Pathology in which restricted subject matter areas in plant pathology would be discussed in great depth. The lectures in this book were those given in the second conference, in the spring of 1968. The charter conference was in 1965 and dealt with "Pathogenesis and Metabolism in Plants."

The University of Wisconsin has had particular strength in plant pathology and has perhaps done as much as any other institution in molding this profession. This has been J. C. Walker's workshop; he has been associated with this department for fifty-four of its fifty-eight years. He has trained more Ph.D.'s, published more papers, and drawn more attention to this department than has any other member. He has been a master builder. His counsel was widely sought and his judgment was as the wisdom of Solomon.

I can think of no other man to whom the agriculture of Wisconsin is so indebted as it is to Walker. Our great vegetable industries of cabbage, peas, beans, beets, carrots, cucumbers, onions, and potatoes have each in turn been sustained and saved by him.

His applied research in Wisconsin has been translated into the profit side of vegetable production throughout the nation. The producers and merchants of fresh vegetables, the American seed industries, and the national processors of vegetable crops have all bestowed honor and recognition upon him because of his service to them.

Perhaps Walker's most lasting contributions will be those in basic research. Specifically, his research in the physiology of disease reactions in plants, which was the subject for the charter J. C. Walker Conference, was of the greatest significance. Corollary to his physiological studies have been those seeking to elucidate the morphological and anatomical aspects of plant disease. The subject treated by this second Conference was particularly compatible with Walker's research interests, and no one was so eminently qualified to deal with the subject as Professor Katherine Esau was.

GLENN S. POUND

Madison, Wisconsin
September, 1968

PREFACE

THE contents of this book represent the material that served as a basis for the series of three lectures given in April and May of 1968 in honor of Professor John Charles Walker of the Department of Plant Pathology, the University of Wisconsin, Madison, Wisconsin. The topical sequence in the book agrees with that of the lectures, but for convenience of the written presentation the subdivision of the contents was somewhat modified. The illustrations at the end of the text are assembled according to the order of first references to them in the body of the text.

The invitation to present the John Charles Walker Lectures provided the motivation to appraise the scope of our present understanding of the morphological aspects of viral pathogenicity, but the lectures were based on original research rather than on a review of the literature. The preparation of the lectures proved a satisfying task and their delivery before the appreciative audience at Madison a great pleasure. Having Professor J. C. Walker in the audience was a special privilege.

I am grateful to the Committee on the John Charles Walker Lectures of the University of Wisconsin for selecting me to give the lectures, and to Professor John E. Mitchell, chairman of the Committee, for the cordial reception he arranged for me at Madison. The hospitality extended to me by the Departments of Plant Pathology and Botany and the stimulating visits in the various departments of the College of Agriculture were greatly appreciated. It is a privilege to have the lectures published by the University of Wisconsin Press, and the splendid cooperation of the officers of the Press in this task is gratefully acknowledged.

In part of my research on viruses I enjoyed a fruitful collaboration of colleagues. The enthusiasm and devotion of Dr. Lynn L. Hoefert made the initial studies on the beet yellows virus a stimulating experience. Dr. James Cronshaw's proficiency in ultrastructural research led to marked technical improvements. The studies pertinent specifically to the lectures were carried out independently but with the most competent assistance of Mr. Robert H. Gill. His technical skills, understanding of the problems involved, and ability to recognize significant features under the electron microscope made his help invaluable. Mrs. Barbara Osterhoff ably assisted with the preparation of material for microscopy.

The research was supported by National Science Foundation grants GB-1523 and GB-5506. The University of California, Santa Barbara, provided space, facilities, and technical assistance. This support, extended to a person in emeritus status, deserves special acknowledgment.

K. E.

Santa Barbara, California
July, 1968

CONTENTS

FOREWORD	v
PREFACE	vii
INTRODUCTION	3
1. INTRACELLULAR FORM OF VIRUS	7
2. RELATION OF VIRUSES TO HOST-CELL COMPONENTS	21
3. VIRUS IN RELATION TO CELL TYPE	31
4. INTERCELLULAR TRANSPORT OF VIRUSES	38
5. RESPONSES OF INFECTED CELLS TO PRESENCE OF VIRUS	49

    Degenerative changes in the chloroplasts, 51
    Degenerative changes in sieve-element plastids, 62
    Phloem degeneration, 64
    The X-material and the P-protein, 70

RETROSPECT	74
BIBLIOGRAPHY	77
ILLUSTRATIONS	89
INDEX	221

VIRUSES IN PLANT HOSTS

INTRODUCTION

ELECTRON microscopy has materially increased our understanding of the relation between viruses and their hosts, the relation that is expressed in the distribution of particles of virus (the virions) within the host-cell protoplast and in the changes wrought in this protoplast by the presence of the virus. The visual aspects of this relation are the theme of this book. Virus-host relationship is discussed on the basis of what is seen in virus-infected cells by means of the electron microscope. In other words, the fundamental approach is morphological. The research that has been carried out in this direction thus far must be characterized as having just begun to contribute to the comprehension of viral pathogenicity.

A meaningful interpretation of virus-host cell relation requires a correlation of morphological, biochemical, and biophysical studies of viruses and the infections they cause. The morphologist has the challenging task of providing the information on the forms the viruses assume in the host cells, their relation to the components of the cell, the paths of movement of virus from cell to cell, and the changes occurring in the host cells in response to the infection. The quality of this information depends to a great extent on the use of appropriate techniques. Although electron microscopy has gone far in showing what the living organism is like in the subcellular realm, further refinements should bring us closer to the molecular level where the definitive explanation of the interaction between the virus and the host must be sought.

The literature pertinent to the theme of the book was recently reviewed (Esau, 1967b). That review and the references published subsequently are utilized as a framework for describing and interpreting personal observations. The discussion of host-virus relations is developed with reference to the two viruses that were used for these observations: the tobacco mosaic virus (TMV) and the beet yellows virus (BYV). The reasons for choosing these particular viruses for research were, first, personal familiarity with their effects on the hosts as seen with the light microscope and, second, the easy access to plants infected with these viruses and grown under conditions precluding their contamination with other viruses. The added advantage in using these viruses was that the appearance of their particles under the electron microscope was well known. Of considerable value is also the circumstance that the two viruses represent two different categories of plant viruses, the mosaic viruses and the yellows viruses. The tobacco mosaic virus shows the characteristic nonselective relation to the host tissues, whereas the beet yellows virus represents the type of virus largely dependent on the phloem tissue for producing a successful infection of the host.

As indicated in the foregoing, the factual material used in the book is largely original, including the illustrations. The information on the two viruses which was published previously (see the bibliography) was re-examined and documented with new illustrations. Some revisions of previous interpretations have resulted from the re-examination. New information gathered since the publication of the latest research paper (Esau and Cronshaw, 1967*b*) constitutes the major part of the text.

The material and methods used in the studies of the two viruses were described in the published papers but are briefly reviewed here for the convenience of the reader. The beet yellows virus was studied in its best-known host *Beta vulgaris* L., the sugar beet. The infected material and control plants were obtained from Dr. C. W. Bennett of the United States Agricultural Research Station at Salinas, California. The test plants were inoculated with one of the virulent strains of beet yellows virus (Bennett's isolate 5, "Brawley strain") by means of the aphid vector, *Myzus persicae* Sulz. The tobacco mosaic virus infections were studied in *Nicotiana tabacum* L. (tobacco) plants systemically infected with the common strain of TMV. The infected plants were supplied by Dr. W. N. Takahashi of the University of California, Berkeley, and by Dr. S. G. Wildman of the University of California, Los Angeles. Control plants were grown in Santa Barbara. For electron microscopy, the beet yellows material and some of the tobacco mosaic material (mainly leaves) were fixed in glutaraldehyde and post-fixed with osmium tetroxide. For most of the tobacco material, glutaraldehyde-paraformaldehyde fixation (Karnovsky, 1965) followed by post-fixation with osmium tetroxide was used. Acetone served for dehydration, Epon epoxy resin for embedding. Sectioning was done with diamond knives on a Porter-Blum MT2 ultramicrotome. Staining with uranyl acetate and lead citrate was carried out on the grids. Observations and photography were made with a Siemens Elmiskop 1.

Material killed in potassium permanganate was used for specific effects in some illustrations. These are appropriately identified in the legends. In three instances, sections of *Cucurbita maxima* Duchesne leaves were used for the illustrations because this material best depicted certain normal structures. Some of the normal structures are illustrated by the use of infected material. These structures did not differ from those in material from healthy plants, but their pictorial presentation happened to be more advantageous in the infected material.

1
INTRACELLULAR FORM
OF VIRUS

ONE of the important advances in plant-virus research effected by the introduction of sectioning into electron microscopy was the identification of virions (virus particles) in situ within the cells of the host. Before the methods of fine sectioning were developed, the ultrastructural studies of viruses were based largely on shadowed preparations made from expressed and purified juice of infected plants. Using such early techniques, Kausche, Pfankuch, and Ruska (1939) achieved the first demonstration of virions with the electron microscope. These virions were particles of tobacco mosaic virus (TMV). Steere and Williams (1953) subsequently showed that the same kind of particles composed the crystalline inclusion bodies which they extracted from TMV-infected *Nicotiana tabacum* (tobacco) cells. The virus was thus identified as such in the host cell. Then, Brandes (1955) demonstrated TMV particles in sectioned material.

Throughout the history of plant-virus research TMV has been receiving the major attention and, correspondingly, has been repeatedly studied with the electron microscope.

One of the especially rewarding aspects of ultrastructural studies of TMV in situ in the host cells is that they enable one to explore the distinction and possible interrelationship between a virus and virus-related material as represented by the characteristic inclusions associated with TMV infections. Cells infected with this virus commonly contain two types of inclusion. One type occurs as crystalline or paracrystalline material in the form of hexagonal plates, needles, spindles, and fibrous structures (Bawden, 1964; Singh and Hildebrandt, 1966; Warmke and Edwardson, 1966b). The other is the amorphous entity called variously vacuolate body, ameboid body, or, most frequently, X-body. The crystalline and paracrystalline material is identified with the virus itself. The X-body is interpreted as being rich in virus-related protein.

Electron microscopy of sectioned material has made it possible to visualize precisely the internal organization of crystalline TMV as previously deduced from X-ray diffraction studies (Wilkins et al., 1950). The three-dimensional hexagonal crystal is composed of layers (Fig. 1) in each of which virus particles occur in parallel alignment, a single particle in depth (Fig. 2). Sections of layers cut transversely to the long axes of the virions show that the particles are arranged in a two-dimensional hexagonal pattern (Fig. 3). According to Warmke and Edwardson (1966b), the individual layers have the capacity for unlimited repetition in two dimensions. Similarly, the stacks of layers in the hexagonal crystal have the capacity for unlimited repetition in three dimensions. The authors envisage the growth of the crystals as beginning with small aggregates of particles appearing in the cytoplasm and subsequently becoming extended into single-layered plates (monolayers), with the virions in parallel arrangement perpendicular to the surface of the plate (Fig. 2). The monolayers then become consolidated into stacked aggregates (Fig. 1).

The ultrastructure of the crystals of tobacco mosaic virus was recognized in some of the early electron-microscope studies of intracellular TMV (e.g., Brandes, 1956; Wehrmeyer, 1959) but was particularly intensively investigated by Warmke and Edwardson (1966a, b). As was found by these authors, a short-time fixation with a weak solution of potassium permanganate is remarkably useful for preserving the crystalline structure (Figs. 1–3). Fixation with glutaraldehyde-osmium tetroxide, however, may also reveal the crystalline structure of TMV (Fig. 4) and is worthy of further testing in this respect because it has the added advantage of giving a better preservation of the host protoplast than is obtainable with potassium permanganate.

In addition to truly crystalline arrangements, the TMV particles form various less orderly aggregations in material prepared for electron microscopy. The crystals themselves may show some variation in organization. In some crystals the particles in the successive layers are accurately aligned; in others, the particles in adjacent layers form a slight angle and produce the familiar herringbone pattern (seen in places in Fig. 1). Some monolayers may aggregate slightly irregularly and entrap components of the host protoplast among them (Fig. 4). In many preparations, units longer than the typical TMV rods form bundles or large aggregates (Fig. 5). The long forms apparently result from end-to-end connections (linear aggregations) of the virions. Warmke and Edwardson (1966b) suggest that such union is a secondary phenomenon and that it occurs in the crystals across the boundaries of the original layers, thus obscuring the limits of the individual virions. The end-to-end union is seemingly stronger than is the lateral attraction between the particles within a given layer so that, when an aggregate is disturbed, it breaks up not into individual particles but into bundles of long rods. This phenomenon explains the striated appearance of the TMV crystals in acid-treated preparations frequently used for light microscopy (Fig. 6). Linear aggregation is observed in many forms of assemblages of virus particles: for example, in prismatic bundles of long rods (Fig. 7), in bulky aggregates of rods disposed in different planes in different portions of the same aggregate (Fig. 8), and in needle-shaped aggregates (Fig. 9).

Warmke and Edwardson (1966b) describe the formation of the needle-shaped crystals as involving end-to-end connection between particles of a crystal and a subsequent splitting of the crystal into bundles of long rods. The needles are then released into the cytoplasm and even into vacuoles. Warmke and Edwardson (1966b) found that in the transverse plane the virus particles in a needle were arranged in hexagonal two-dimensional regularity, but the authors remained uncertain about the alignment of particles in longitudinal direction (see also Bawden, 1964). Linear aggregations may produce extremely long fibrous structures, sometimes extending through the entire cell in straight or looped condition (Singh and Hildebrandt, 1966). Warmke and Edwardson's (1966b) interpretation that the needle crystals and other elongated forms are secondary aggregation products resulting from a breakdown of the original crystalline inclusions is supported by the frequent occurrence of such forms in older leaves of plants with long-standing infections. Transformations of the hexagonal crystals into elongated paracrystalline forms may be induced artificially in vitro and within the cells containing them, but Warmke and Edwardson (1966b) think that the phenomenon probably occurs naturally as well.

Purcifull and Edwardson (1967) found end-to-end connections of virus particles in extracts from pumpkin leaves infected with watermelon mosaic virus and thought that such aggregations occurred also in vivo. Thus, one may expect considerable natural variation in the form of viral aggregates in sectioned material. At the same time, the possibility that some of the forms result from displacements induced by the methods of preparation cannot be ignored.

The true crystals reveal the characteristic shape and length of the TMV particle. It is a stiff rod approximately 3000 A long and 150 A in diameter. Studies with a variety of methods have led to the concept that in the TMV particle both the subunits of the coat protein and the single ribonucleic-acid thread are helically arranged. The helix forms a tube with a canal 40 A in diameter. The ribonucleic-acid (RNA) strand is embedded in the protein 20 A from the edge of the canal. Electron-microscope views of transections of virions in stained preparations provide some evidence of the double structure of the TMV rod: the region next to the canal appears more electron opaque than does the peripheral region (Fig. 10). Such views are interpreted as indicating the differentiation of the particle into the ribonucleoid region near the canal and the protein-coat region along the periphery (e.g., Milne, 1967b). Gardner (1967) observed a similar differentiation in the particles of the barley stripe mosaic virus, and he was able to enhance the differentiation by modifications in the staining procedure.

When the TMV particles are seen in longitudinal views the differentiation into the nucleic-acid region and the coat region is commonly not readily visible because of the poor stainability of the coat (Fig. 11). Consequently the particle appears thinner than would be expected from the 150 A value of its diameter (Fig. 12); it is mainly the electron-opaque, RNA-containing core that is discernible (e.g., Kolehmainen, Zech, and Wettstein, 1965; Milne, 1967b). The TMV particles in Figures 12 (lower part of aggregate) and 13 (middle of aggregate) deviate from this appearance. A component consisting of subunits in helical arrangement seems to cover the narrow cores. If this component constitutes the protein coat, its strong stainability is exceptional and cannot be readily explained. Arnott and Smith (1968) observed a rather similar deeply stained helical component in a virus of the mosaic type affecting *Lantana horrida*. The authors assumed, however, that the helix was part of the core of the particle and that it was later covered with a coat. According to Figures 12 and 13, in the TMV particles the helix appears to be outside the core.

The interpretation of the nature of the X-bodies (Figs. 14 and 15) is more problematic than that of the virus crystals. With the introduction of fixatives that preserve proteins and nucleoproteins, aggregates of tubules began to be recognized in sections of TMV-infected tissues. Since this component was found to be associated with virus particles it was immediately suspected of having some relation to the synthesis of the virus (Kolehmainen et al., 1965; Milne, 1966, 1967b; Shalla, 1964), possibly serving as coat protein. Such interpretation is not out of line with the fact that the juice expressed from TMV-infected plants contains not only complete virus particles but also non-infectious protein similar to that forming the coat of the virus particle. This protein was named X-protein by its discoverers, Takahashi and Ishii (1953; see also Knight, 1963, p. 117). Bawden (1964, pp. 49–58), on the other hand, postulated that, within the cells, the virus-related protein occurred in the X-bodies. By comparing light-microscope views with electron micrographs of TMV-infected cells Esau and Cronshaw (1967a) identified the aggregates of tubules associated with the virus in the host cells as the X-bodies (Figs. 14 and 15). Possibly the X-bodies contain surplus protein similar to the noninfectious X-protein of Takahashi and Ishii (1953). As a component of the X-bodies (so named by Goldstein in 1926) the protein tubules could be called X-protein; but to avoid the implication that the tubular material of the X-bodies is identical with the X-protein of Takahashi and Ishii (1953) it is referred to in the following as X-material, X-tubules, or X-component. The concept that the X-component is surplus viral protein, however, is attractive and need not be taken to mean that this protein may not be utilized, at least in part, to form the coat of virus particles.

The aggregates formed by the X-tubules are morphologically variable. In many X-bodies the tubules occur in close association with one another in groups, mostly in multiples of three. Such associations are clearly revealed in transections of the groups (Fig. 16, *arrows*). From the sides the groups appear as flexuous bands somewhat variable in width probably because of differences between planes of sectioning of the triplets of tubules. In other aggregates the tubules are not in groups and appear to be less flexuous than the bands of tubules. Esau and Cronshaw (1967b) have suggested that the variations in the tubule aggregates reflect a developmental modification of the X-material; additional observations have strengthened and expanded this concept. In the earliest stages detected so far, the X-material appears as aggregates of granules (Fig. 17). Somewhat older aggregates consist of mixtures of granules and of flexuous bands; that is, composites of tubules (Fig. 18). It seems plausible to assume that the granular aggregates are composed of protein subunits which become polymerized into tubules. The number of tubule complexes increases, host-cell components become included among the complexes, and the aggregates assume the familiar form and size of the X-bodies that are detectable with the light microscope (Fig. 14). Ribosomes and cisternae of endoplasmic reticulum are the chief host-cell components represented in the X-body (Fig. 16). The bodies may be in contact with nuclei, chloroplasts, mitochondria, and dictyosomes. Virus particles frequently occur in the X-bodies, dispersed or in small aggregates without bounding membranes (Fig. 15). Older X-bodies contain vacuolelike cavities enclosing some host-cell components or unidentifiable degenerated material. The X-body itself is not bound by a limiting membrane. It has the appearance of a region in the cytoplasm in which protein synthesis, carried out by host-cell components, is concentrated. One may assume that the X-bodies are regions of virus-directed protein synthesis, including the polymerization of subunits into tubules.

The X-tubules do not remain in the closely knit complexes of three but separate from one another and become individual tubules assembled into aggregates of various sizes. The separation of tubules is apparently a gradual process, for one encounters aggregates in which triplets of tubules are dispersed among single tubules (Fig. 19). In the final stage, the single tubules appear in more or less orderly arrangements in homogeneous aggregates free of host-cell components (Fig. 20). The tubules are now rather straight and vary in length (Figs. 20–22). Conceivably, the longer tubules are formed through end-to-end connections of shorter tubules. The variation in tubule length in Figure 21 may be interpreted in this sense.

The separation of the members of triplets of tubules begins in the still circumscribed X-bodies, but later the aggregates of disconnected tubules become dispersed and may be found in isolation from others within the host cytoplasm (Fig. 22). The timing in the dispersal of the aggregates is apparently variable. Sometimes the aggregates of individual tubules occur together in a body that has all the characteristics of an X-body except that the tubules do not form bands of triplets (Fig. 21). The existence of X-bodies of this type supports the concept of developmental relation between the earlier flexuous bands of tubules and the later, more rigid single tubules assembled into more or less orderly aggregates. These single tubules resemble superficially the microtubules which constitute characteristic components of the parietal cytoplasm in cells with growing walls (Fig. 22). Transectional views of microtubules (Fig. 17), however, indicate that they have a much narrower electron opaque peripheral region than do the X-tubules.

It is pertinent to recall at this point that the X-protein discovered by Takahashi and Ishii (1953) consisted of approximately spherical subunits which, in purified state, were polymerized into rod-shaped particles resembling those of TMV. One might draw a parallel between this sequence and the formation of tubes from the apparently granular material seen at the initiation of X-body development (Fig. 17). The great length eventually attained by the X-tubules need not necessarily indicate a difference from the coat protein. According to some examples in the literature, the coat protein not associated with nucleic acid may assume a form that strikingly deviates from that which it shows as a component of a complete virus particle. Bancroft, Hills, and Markham (1967) found that the protein of the spherical cowpea chlorotic mottle virus can be made to form long tubules in vitro, and Hitchborn and Hills (1967) discovered tubular protein units associated with the spherical turnip yellow mosaic virus in vivo. Bancroft, Hills, and Markham (1967) characterized the anomalous products assembled from viral components as mistakes resulting from the lack of association with intact nucleic acid. There is, on the other hand, the indication that the nucleic acid does not necessarily direct the form assumed by the protein during the assembly of the virus particle. In hybridization experiments with viruses Hiebert, Bancroft, and Bracker (1968) found that hybrid particles combining TMV-RNA with the coat protein of the spherical cowpea chlorotic mottle virus were spherical; that is, the particles assumed the form of the donor of the protein coat.

A more outstanding difficulty in the interpretation of the nature of the X-material is the difference in width between the complete virus particle and the X-tubule as seen in situ in the host cells. The X-tubules are almost twice as wide as the virions (Fig. 20). The TMV particle is on the average 150 A wide, whereas the X-tubules, as measured by Esau and Cronshaw (1967b), are approximately 280 A wide. The X-material and the coat protein of the complete particle also differ in stainability. After the same kind of processing, the X-tubules appear more electron opaque than does the coat of the complete particle, especially if the two are compared in longitudinal sections. In Figure 20 the X-tubules appear almost as dark as the ribonucleoid cores of the virus particles.

Perhaps the X-material assumes a form and structure deviating from that of the coat protein when it constitutes a by-product of viral synthesis. Should it later become the source of coat protein, it could, conceivably, become depolymerized into the original subunits before participating in the assembly of the virus particle. Zech (1960) visualized a decrease in hydration during the assembly of the particle, but the voluminous and flexible entities described by Zech as the intermediate product of the assembly may have been not RNA-containing material but the flexuous bands of tubules found in the early form of X-body. Kolehmainen, Zech, and Wettstein (1965, p. 80) made a similar interpretation of Zech's earlier observations except that they regarded the bands of X-tubules as "loosely spiralized voluminous forms of TMV particles."

The forms assumed by the beet yellows virus (BYV) in the cells of *Beta vulgaris* (sugar beet) have also been studied comparatively with the light and the electron microscopes. Cells infected with the virus may develop conspicuous inclusions, of which some are homogeneous, others are banded, and still others are aggregates of fibrous material (Esau, 1960a, b). As seen with the electron microscope all of these types of inclusions consist of virus particles (Cronshaw, Hoefert, and Esau, 1966; Esau, Cronshaw, and Hoefert, 1966, 1967). No structures resembling X-bodies of the type found in TMV infections have been seen in sugar beets infected with BYV. Figure 23 illustrates a banded type of inclusion at low power of magnification and Figure 24 reveals that the banding is a reflection of the layered arrangement of virus particles. The alignment of the virions within the layers of the BYV inclusions is less orderly than it is in the TMV crystals possibly because the BYV particle is flexuous and much longer than the tobacco mosaic virion. Nevertheless, transectional views of aggregates of the beet yellows virions may show considerable order in their distribution (Fig. 25). The banded aggregates are thus at least paracrystalline. Nonbanded inclusions show, longitudinally, a random massing of virus particles (Fig. 26) with some tendency toward a two-dimensional ordering in the transverse plane (Fig. 27). Besides the two kinds of circumscribed inclusions of virions, small bundles and longer strands, as well as formless masses of virions, are interspersed with host-cell components in many cells (e.g., Fig.

28). Longitudinally extended aggregates, which may consist of particles connected end-to-end, constitute the structural elements of the fibrils visible with the light microscope.

Having a length of 12,500 A, the beet yellows virus particle was considered for some time to be the longest known plant virion (Brandes and Wetter, 1959). According to Kitajima and co-workers (1964), however, the virion of the citrus tristeza disease is longer (20,000 A). The BYV particle is approximately 100 A wide and has a central canal 30–40 A in diameter. It is helical in organization but the helix appears to be somewhat more loose than that of the TMV particle (Russell and Bell, 1963).

Crystalline and random aggregations of virions have been seen in many different viral infections (cf. Esau, 1967b), including those induced by isometric (spherical or polyhedral) and anisometric (rods, flexible threads, bacilliform types) viruses (classification by Bawden, 1964, p. 205). Arnott and Smith (1967) added another virus to the group inducing the formation of inclusions which consist of curved plates and resemble pinwheels in transections (cf. Edwardson, 1966). The authors suggest that these inclusions may be composed of viral protein rather than of virus particles. Long flexuous rods, hexagonal in transection, were suspected of being the virus. A distinction between rod-shaped virus particles and inclusions composed of curved plates was made also with regard to watermelon mosaic and tobacco etch viruses (Edwardson, Purcifull, and Christie, 1968; Purcifull and Edwardson, 1967). De Zoeten and Schlegel (1967a) observed a dense packing of the spherical broadbean mottle virus particles in the host cells and interpreted the crystallization of the virus as a result of saturation of the host cytoplasm with virus. Russo,

Martelli, and Quacquarelli (1968) described the cytoplasmic distribution of the spherical mottled crinkle virus of artichoke as in part random, in part crystalline. Weintraub, Ragetli, and John (1967) proposed that morphology of the TMV particle aggregates depended on whether the infection was systemic or localized. In the latter instance they found the aggregates to be small, appearing as single rows of about 25 particles or occasionally as double rows. In systemic infections the accumulations were massive and were either in crystalline or in random arrangements. Actually, as illustrated in this book, considerable variation in the size of virus aggregates is found in the systemically TMV-infected *Nicotiana tabacum*. Fujisawa, Hayashi, and Matsui (1967) studied mixed infections involving two serologically unrelated viruses which were inoculated into the same plant of *Nicotiana tabacum*. Inclusions characteristic of infections with the two viruses, the tobacco mosaic virus and the tobacco etch virus, occurred in the same cell when the inoculations of the two viruses were made simultaneously. In fact, the components of the two kinds of inclusions were intermixed.

A notable achievement in the survey of form of viruses in host cells was the recognition of the same type of particles, in similar arrangements, in the cells of both the host plant and the insect vector. Instructive examples of such relation are found in the wound tumor virus (Shikata and Maramorosch, 1965, 1966a, 1967: vector, the leaf hopper *Agallia constricta*; host plants, *Melilotus officinalis, Rumex acetosa*, and *Trifolium incarnatum*), pea enation mosaic virus (Shikata, Maramorosch, and Granados, 1966: vector, the aphid *Acyrtosiphon pisum*; host plant, *Pisum sativum*), and rice dwarf virus (Fukushi, Shikata, and Kimura, 1962; Fukushi and Shikata, 1963: vector, the leaf hopper *Nephotettix apicalis* var. *cincticeps;* host plant, *Oryza sativa*). All three viruses belong to the group with isometric particles.

The foregoing discussion indicates that the inclusions associated with viral infections may assume diverse forms and various internal organizations. Some inclusions consist of virus particles, others contain mainly virus-related protein. Host-cell components may be associated with the viral inclusions, probably because of their involvement in processes of viral synthesis. In final stages of development, however, virus particles and viral by-products may form homogeneous aggregates free of host-cell components. Conceivably, inclusions may also consist of aggregations of degenerated host material combined with viral products (see Chapter 5). The virus particles as such often assume ordered arrangement in crystalline or paracrystalline aggregates. They may be extremely numerous in a given cell, even to the point of saturating the cytoplasm. Inclusions of virus-related material have been recorded in some viral infections and not in others. This material also assumes diverse forms in its arrangement, some being rather specific for certain groups of viruses. A proper interpretation of the virus-related material in terms of its possible role in virus synthesis is still to be made. The description and classification of the forms of viral inclusions, as seen with the electron microscope, must be made with caution because some of them may result from responses to preparatory treatments, or at least may become modified by such treatments.

2

RELATION OF VIRUSES TO HOST-CELL COMPONENTS

VISUAL localization of virus particles in the host cell is of considerable interest with regard to the question where, precisely, in the cell the processes of viral synthesis are taking place. A considerable uniformity of opinion has been reached in this regard for the bacterial and the animal viruses, but the views on the sites of virus multiplication in plant cells are varied and to some extent contradictory. According to a common concept, based mainly on studies of the tobacco mosaic virus, the virion entering the host cell first becomes separated into two moieties, the nucleic acid and the protein; second, the nucleic acid undergoes duplication and serves as a code for the formation of viral protein; and third, the new nucleic acid and the new protein are combined, or assembled, into complete virus particles (Luria, 1958; Mundry, 1963; Sanders, 1964; Schramm, 1961; Shaw, 1967). (Commoner, 1959, however, holds the contrary view that the nucleic acid and the protein are not formed separately but are joined at all times and that, in TMV, the particle grows by accretion at one end.)

The proponents of the concept of separate origin of viral RNA and protein disagree regarding the location within the cell of the processes of synthesis of the two components and their assembly into complete particles. One of the prominent views, which is supported by a considerable amount of experimental data, is that the viral RNA is synthesized in the nucleus and is then released into the cytoplasm where it serves as a code for the formation of viral protein (Schlegel, Smith, and de Zoeten, 1967; Schramm, 1961; Wettstein and Zech, 1962). According to a contrasting view, concerning specifically TMV, the nucleic acid and the protein are both synthesized in the nucleus and are also assembled there into complete particles (Reddi, 1966). Still another concept refers to the chloroplasts as the possible sites of protein synthesis or of the assembly of TMV particles (Zaitlin and Boardman, 1958). Finally, synthesis in the cytoplasm has been suggested for some viruses (Schlegel et al., 1967, p. 242).

Although the presence of complete virus particles within one or another component of the protoplast does not necessarily indicate that the virus originates in that component, it cannot be ignored in the discussions of the sites of virus multiplication. Moreover, if the type of material described as X-component for TMV infections is identical with the coat protein, its distribution may indicate specifically the sites of synthesis of viral protein and possibly also those of the assembly of the particles. The presently available information indicates that viruses differ in their relation to the host-cell components. Some appear to be restricted to the cytoplasm, at least in their complete form; others occur also in one or another type of organelle.

The occurrence of virions (or of presumed virions) in the cytoplasm has been recorded for numerous viruses (cf. Esau, 1967b; and additional observations by Arnott and

Smith, 1967, 1968; De Zoeten and Schlegel, 1967a; Gerola and Bassi, 1966; Lee, 1967; Milne, 1967a; Russo et al., 1968; Shikata and Maramorosch, 1966a; Shikata et al., 1966). Figure 28 illustrates beet yellows virus particles in the cytoplasm of a cell from a sugar beet leaf and Figures 7, 8, 13, 15, 21, and 22 show aggregates of tobacco mosaic virus and virus-related X-material in the cytoplasm of cells of tobacco leaf. Tobacco mosaic virions may be seen in vacuoles (e.g., Fig. 15), but such distribution probably results from an accidental disruption of the tonoplast of the cell. It is not uncommon to observe breaks in the tonoplast with virus particles apparently fixed as they were being spilled into the vacuole. In Figure 29, for example, three of the four vacuoles in the cell in the center of the picture are free of virus, but the large vacuole to the right contains numerous TMV particles which presumably egressed from the cytoplasm through the break in the tonoplast indicated by the arrow. The X-material in TMV infections was seen only in the cytoplasm. Beet yellows virus particles were occasionally encountered in vacuoles.

Whether viruses naturally occur in vacuoles is somewhat controversial (cf. Esau, 1967b, p. 53). Warmke and Edwardson (1966b) mention that the paracrystalline needles of TMV may enter the vacuoles. Russo, Martelli, and Quacquarelli (1968) visualize a kind of inverted pinocytosis responsible for the release of the spherical particles of the mottled crinkle virus of artichoke into the vacuole. The authors describe the formation of cytoplasmic protuberances directed toward the vacuole, the filling of the protuberances with virions, their severance from the parietal cytoplasm, and their eventual disruption. Some authors use the crystallization of virus particles in vacuoles as evidence that their occurrence in that position is a normal phenomenon (Esau and Cronshaw, 1967a; Russo et al., 1968). It is well known, however, that viruses may be made to crystallize in various media in vitro (Bawden, 1964, chapter 11).

Intrusion into vacuoles of cytoplasmic protuberances containing virions—the phenomenon used by Russo, Martelli, and Quacquarelli (1968) to explain the entry of virus particles into vacuoles—has been recorded in more than one kind of viral infection (e.g., Gardner, 1967; Weintraub et al., 1967). Russo and co-workers (1968) assume that the formation of the protuberances is not related to viral multiplication. According to these authors, the virions enter the protrusions after these are initiated. In cells infected with tobacco mosaic and beet yellows viruses, the evagination of tonoplast-bound cytoplasm, which is frequently almost saturated with virions, gives the impression that the phenomenon is induced by space requirement for the increasing amount of virus or for the growth of a crystal. The virus aggregate is sometimes accommodated within the parietal cytoplasm in a bulge resembling those associated with the larger organelles (Fig. 30). The protrusion containing the virus may also assume the form of a pouch or fingerlike extension deeply intruding into the vacuole (Figs. 29 and 31). A virus aggregate sometimes forms a bulge in a cytoplasmic strand traversing the vacuole (Fig. 32). Transverse sections through the protrusions and bulges frequently have circular outlines (Figs. 33 and 34). The virus in a protrusion is associated with small amounts of cytoplasmic components and is surrounded by the tonoplast. No decisive evidence has been found that a virus aggregate growing in a protrusion may burst through the tonoplast. The arrangement of virus particles in a protrusion is similar to that elsewhere in the cytoplasm but sometimes the particles assume an orientation related to the shape of the protrusion. In Figure 34, for example, the particles of beet yellows virus are concentrically arranged parallel with the circumference of the protrusion.

Tobacco mosaic virus particles were encountered in intercellular spaces, sometimes in massive aggregates of crystalline structure (Fig. 35). It is likely that virus may become spilled into intercellular spaces when the plant is cut. If this is the only method of invasion of the spaces by a virus, one must assume that the entry is rapidly followed by crystallization of the virus.

With regard to the organelles, virions were most frequently recorded in nuclei. In addition to the examples of occurrence of complete virus particles in nuclei reviewed by Esau (1967b, p. 56), several other pertinent observations should be mentioned. Gardner (1967) demonstrated particles of barley stripe mosaic virus in host nuclei. Shikata and Maramorosch (1966b) found that in infections with the pea enation mosaic virus the nuclei became filled with the isometric virions and were eventually destroyed. Lee (1967) observed the bacilliform particles of wheat striate mosaic virus within the nucleus and between the two membranes of the nuclear envelope. Arnott and Smith (1968) found inclusions of a mosaic type of virus of *Lantana horrida* in the nucleus. In Figure 28, beet yellows virions are seen in great abundance in the cytoplasm of a beet-leaf cell but there is also a conspicuous aggregate of virions in the nucleus. In Figure 36 the nucleus of a young mesophyll cell of tobacco includes a small aggregate of TMV particles. X-material of TMV infections has not been encountered in nuclei in electron-microscope studies, but the search must continue because Woods and Eck (1948) reported finding X-bodies with the light microscope in nuclei of tobacco infected with a certain strain of tobacco mosaic virus.

In their review of sites of virus synthesis in host cells, Schlegel, Smith, and de Zoeten (1967) stress the strength of the evidence that virus, at least that of the tobacco mosaic, is synthesized in the nucleus, although this synthesis may not involve the protein component of the particle. If the protein is synthesized in the cytoplasm, possibly the assembly of the particles would occur here also. Under such circumstances, the presence of complete virus particles in the nucleus could mean that such particles enter the nucleus from the cytoplasm. The other possibility would be that the coat protein migrates from the cytoplasm into the nucleus and combines there with the infectious RNA into complete particles (but see Hirai and Wildman, 1967). Virus particles or protein subunits could, possibly, pass through the pores of the nuclear envelope, but the proposal (Schlegel et al., 1967) that the "channels" described by Wettstein and Zech (1962) in the multilobed nuclei of infected cells serve for entry of virus into the nucleus does not seem to be useful. The "channels," formed as a result of convolutions of the nucleus, are lined with the nuclear envelope and provide no openings into the interior of the nucleus other than the usual pores. Figure 37 clearly illustrates this relationship. The much lobed nucleus, a portion of which appears to the left in the picture, has numerous pores in an otherwise continuous envelope. Aggregates of virus occur in the cytoplasm filling the invaginations between the nuclear lobes but are isolated from the nuclear interior by the envelope. Incidentally, the degree to which the nuclear pores are open is by no means a settled matter (cf. Sitte, 1965, p. 62). Frequently the pores appear to be obstructed by a membrane with an electron-opaque central dot (Fig. 38). Feldherr's (1962, 1965) studies on the passage of gold particles into nuclei of amebas suggest that the actual opening may be smaller than the dot. If the observations of Singh and Hildebrandt (1966) are correct, TMV in the form of needles and angular crystals may penetrate the nucleus during cyclosis. In fact, the authors report that crystals move into and out of the nucleus. Because of the size of the crystals, their movement in and out of the nucleus would presumably involve rupture of the nuclear envelope. This aspect of nucleus-virus relation has not been studied with the electron microscope.

A particularly attractive hypothesis to explain the presence of complete virus particles in nuclei would be to assume that virus or virus-specific protein becomes included in the nucleus during mitosis. As is well known, at the beginning of mitosis the nuclear envelope breaks down. Its components, interpreted as cisternae of endoplasmic reticulum, and probably other ER, form conspicuous accumulations at the poles of the mitotic spindle (Fig. 39). At this stage the cytoplasm and the nucleoplasm are not delimited from one another. In dividing cells infected with TMV, aggregates of virus particles can be easily found close to the chromosomes (Figs. 39 and 40), that is, in the region from which cell organelles are removed during mitosis. One could argue, of course, that the virus occurring in such position was present in the nucleus before the envelope was removed. In Figure 41, however, the two large aggregates of virus have undoubtedly arisen in the cytoplasm, yet one of them is in contact with a chromosome. It may also be significant that virus aggregates associated with chromosomes of dividing nuclei are encountered more frequently than are those occurring in intact nuclei.

The presence of virus particles in chloroplasts is apparently a rare phenomenon: it has been recorded unequivocally only for TMV (Fig. 42; Esau and Cronshaw, 1967*a*). At the same time, in TMV infections chloroplasts with virions may be encountered much more frequently than virus-containing nuclei. In one particular mesophyll cell, five chloroplasts among seven present in a single field of view contained TMV particles. In TMV-infected tobacco, virions were recorded also in immature chloroplasts (Fig. 43), in young plastids of immature tracheary elements (Fig. 44), in chloroplasts of guard cells, and in leucoplasts of glandular hair cells and crystalliferous cells (Fig. 45). Of particular interest is the relation of virus to sieve-element plastids, which typically, in dicotyledons, show little internal differentiation. Relatively young sieve-element plastids of *Nicotiana tabacum* show the characteristic double-membraned envelope, a somewhat dense stroma, extremely sparse internal membranes, and prominent starch granules (Fig. 46). (This starch stains red with iodine.) At maturity, the plastid shows little change except that the matrix becomes less dense (Fig. 47). In TMV-infected material, both immature and mature sieve-element plastids may contain virus particles (Figs. 48 and 49). The distribution of TMV particles throughout the plastidome of the host, including immature and little differentiated types of plastid, indicates that their occurrence in the chloroplasts specifically is not related to photosynthetic activity. The evidence that virus multiplication can occur in chlorophyll-free tissue is in accord with this assumption. As Schlegel, Smith, and de Zoeten (1967) have pointed out, no clear suggestion has been made about the specific role of chloroplasts in virus synthesis.

In certain isometric viruses, particle aggregates are enclosed in membranes interpreted to be those of the endoplasmic reticulum (cf. Esau, 1967b, pp. 54–55). The isometric particles of viruses inducing tumors have been seen in rows within straight tubes or other forms of membranous containers (maize rough dwarf virus, Gerola and Bassi, 1966; wound tumor virus, Shikata and Maramorosch,

1966a). In certain other virus infections virions were located between the two membranes of the nuclear envelope (cf. Esau, 1967b, p. 55; also Lee, 1967). Virions have not been seen in dictyosomes. Russo, Martelli, and Quacquarelli (1968) are apparently the first authors to report plant-virus particles in mitochondria.

Schlegel, Smith, and de Zoeten (1967) suggested that viruses may differ with regard to the sites of virus-directed synthesis. De Zoeten and Schlegel (1967b) considered the possibility that a relationship may exist between the shape of the virus and the site of its synthesis. In a comparative study, these authors found evidence of synthesis in the nucleus for certain rod-shaped viruses but not for the spherical viruses used. In the latter type of virus, virus-directed synthesis appeared to be located in the cytoplasm. Even strains of the same virus may show differences in their activities within the host cells (Schlegel et al., 1967).

A distinct phenomenon, related by certain authors to virus multiplication, was observed in two tumor-inducing viral infections. Gerola and Bassi (1966) reported that immature virions of maize rough dwarf virus, that is, virions lacking the light region surrounding the dark core, are generally found in dense masses of fibrils, the viroplasm. Such material was formerly known only in animal viroses. The authors suggested that possibly the viroplasm produces the protein shell of the particle. In older tumors the viroplasmic fibrils become more tightly packed, whereas the complete virus particles form crystalline aggregates. The other reference to viroplasm was made by Shikata and Maramorosch (1967). In their search for foci of assembly of the wound tumor virus in several host plants and in the insect vector, *Agallia constricta*, these authors observed the so-called viroplasm in both the plant and the insect. They described the viroplasm as tightly packed aggregates of filamentous and granular elements located in the cytoplasm. Particles of wound tumor virus were associated with the viroplasm, first as single units at the periphery of the viroplasmic aggregate, later in larger numbers throughout the viroplasm. As judged by the illustrations, the viroplasm described for the two tumor-inducing viral infections has a considerable resemblance to the dense aggregates of granules interpreted as incipient tobacco mosaic X-bodies (Fig. 17).

3

VIRUS IN RELATION TO CELL TYPE

STUDIES dealing with transmission of viruses from infected to noninfected plants, their transport within the body of the host, and the character and development of external and internal symptoms in the infected plant have clearly demonstrated that viruses differ in their relation to the tissues of the host. Some viruses appear to be adapted to specific tissues, others are unrestricted in their invasion of the host body. The more restricted viruses are mainly those that depend on the phloem tissue for a successful establishment of infection, but occasional viruses utilize the xylem tissue equally successfully. The various tissue relations of viruses as revealed by anatomic and physiologic studies have been previously reviewed (Esau, 1961, 1967b). Briefly characterized, the relations are the following: phloem-dependent viruses are introduced by the vector into the phloem, possibly directly into sieve tubes; the first internal symptoms of infection by such viruses are detectable in cells next to the sieve elements, an indication that the virus is carried in the food conduit to noninfected parts of the plant; the nonrestricted viruses induce infection when introduced into any type of living cell but utilize the food conduit in their long-distance transport through the plant. The expression of these relations at the ultrastructural level is yet to be fully explored, but some information on the subject is available for the viruses of tobacco mosaic and beet yellows diseases. Tobacco mosaic virus is representative of the type of virus that is nonrestricted in its methods of entry into the host and its spread during the establishment of systemic infection. The beet yellows virus exemplifies a virus largely concentrated in the phloem tissue although not as strictly as some other phloem-dependent viruses; it spreads into the ground tissue after the initial invasion of the phloem. In the following, the two viruses are examined with regard to the kinds of host cells in which they have been recognized as virions or as virus-related material or as both. Although this examination is based on surveys of considerable extent, it cannot be regarded as definitive. The use of extremely small fragments of tissues for examination under the electron microscope may lead to negative results simply because the cells in question are inadequately represented in the sample.

The distribution of TMV inclusions visible with a light microscope was comprehensively surveyed, especially by Goldstein (1926). Electron microscopy has confirmed these early findings with regard to the conclusion that tobacco mosaic virus invades all kinds of cells. Virus particles may be found in unspecialized parenchyma cells and in more or less highly specialized cells. Particles are as well represented in immature, still dividing cells, as they are in fully differentiated, mature cells. Epidermal hair cells have served repeatedly for the study of development of viral inclusions in living state, and the electron microscope readily reveals aggregates of virus particles in such cells. For a time the occurrence of virus in guard cells of a stoma was under question; it was proposed that guard cells developed no TMV inclusions because they had no plasmodesmata in their walls (Sheffield, 1936). Later, both TMV inclusions and plasmodesmata were identified in the guard cells of tobacco (Esau, 1941). Electron microscopy demonstrates TMV inclusions in guard cells (Fig. 50) and plasmodesmata in their walls (Fig. 51). Virus also occurs in the still meristematic precursors of the guard cells and is passed from cell to cell during division. Virus is found in cells with peculiar cytoplasmic features such as the crystal cells. In the crystalliferous cell in Figure 52, a virus aggregate is surrounded by a voluminous array of rough endoplasmic-reticulum cisternae frequently present in such cells. Tobacco mosaic virus is common in mesophyll cells (e.g., Figs. 30–32). X-material also was seen in the various cells mentioned above.

The possible presence of virus particles in the main conducting cells, the sieve elements and the tracheary elements, is of particular interest with regard to translocation of viruses and the character of virus-directed synthetic processes in specialized cells. At maturity, the sieve elements have considerably modified protoplasts and the tracheary elements are entirely devoid of protoplasts. These features may facilitate the movement of viruses through the conduits but whether they also provide a favorable environment for the survival of the virus is problematic. Previous research (Esau and Cronshaw, 1967a) and additional observations reveal the following relations of TMV to the sieve elements. Virus particles occur in mature sieve elements as would be expected from the evidence that TMV is translocated in the phloem. Frequently, most of the virions seen in a given section occur in an aggregate apparently surrounded by a membrane, which may be in a state of degeneration (Esau, 1967b, Fig. 2). Figures 53 and 54 show that cisternae of endoplasmic reticulum of the type characteristic of mature sieve elements may serve as a bounding membrane of viral inclusions. The confinement of the virus to aggregates bound by endoplasmic reticulum does not seem to be favorable for efficient transport of the virus through the sieve tube. Further search revealed, however, that particles may be freely dispersed in the sieve element (Fig. 55) and, in fact, may be oriented in a way suggesting a vertical flow along the length of the cell (Fig. 56). The spread of virus in the lumen of the sieve element is in harmony with the concept that in the mature conduit the cytoplasm is not delimited by a tonoplast. In contrast, in parenchyma cells the tonoplast confines the virus to the cytoplasmic layer except when it is, presumably accidentally, disrupted (Fig. 29). Virus particles may also occur in sieve-element plastids (Figs. 48 and 49).

The presence of complete virus particles in mature sieve elements raises the question of the source of this virus. In its cell-to-cell movement in infected leaf tissue the virus eventually reaches the veins and could enter the sieve element from adjacent parenchyma cells through plasmodesmata connecting the two kinds of cell. At the same time, the virus could penetrate also into immature sieve elements and be retained in these cells until their differentiation is completed and the contents become mobile. Immature sieve elements containing tobacco mosaic virus particles are readily observed in infected plants. In Figure 57, for example, a TMV aggregate occurs next to a P-protein body (slime body) in a young sieve element. The density of this body indicates an early stage of differentiation in the sieve element, a stage at which the nucleus is still intact and a tonoplast-delimited vacuole is present. Virus may be observed in sieve elements in various stages of differentiation. In Figure 58 it is associated with partly dispersed P-protein, a stage of development succeeding the one shown in Figure 57. In the former survey (Esau and Cronshaw, 1967a, b) only complete TMV particles were seen in the sieve elements. Subsequent observations showed that X-material may be present in these cells as well. Figure 59 illustrates portions of two young sieve elements, in one of which complete virus particles and an X-body occur next to a P-protein body. The X-body is at the stage when the tubules are assembled into flexuous bands in groups of three. The later form of the X-material, the form distinguished by homogeneous aggregations of individual tubules, was also found in differentiating sieve elements. So far, no mature sieve elements containing X-material have been recorded.

Tobacco mosaic virus is not known to be naturally transported in the xylem. Nevertheless, mature tracheary elements frequently contain particles of this virus. In Figure 60, portions of two contiguous tracheary elements are seen filled with TMV particles. In one cell the particles are in vertical alignment, in the other, mostly in random distribution. Figure 60 also shows the now familiar phenomenon of disorganization of the primary walls in primary tracheary elements (cf. Esau, 1967a). Only fine fibrillar material has remained in the place of the formerly solid primary wall. Nevertheless, the virus particles did not appear to have been passing laterally between the two tracheary elements.

The question about the source of virus raised with regard to the mature sieve elements may be repeated with regard to the mature tracheary elements. Xylem-parenchyma cells frequently contain TMV particles in infected plants, but the passage of virus from a living parenchyma cell into a mature nonliving tracheary element is difficult to visualize. If, however, the virus present in a mature tracheary element is there because it entered the immature cell, one must assume that such tracheary element continued to differentiate in the presence of virus and that the disintegration of its protoplast did not affect the virus. Young tracheary cells containing virus particles are readily encountered (Fig. 61) and such cells may contain X-material as well. In Figure 62 the X-body in the immature tracheary element is at the stage of development when the tubules are still in composites of three, but the material seems to be in a state of disintegration and is associated with inflated endoplasmic-reticulum cisternae. The nucleus, which includes small enclaves with virus particles, is also partly degenerated. Since the secondary wall appears to be fully developed in the element, its protoplast was probably undergoing normal breakdown. Possibly the X-material was being destroyed at the same time.

The information regarding the types of cells invaded by the beet yellows virus is less complete than is that for the tobacco mosaic virus. Viral inclusions or dispersed virus particles have been observed in epidermal cells, including guard cells, in various parenchyma cells (Figs. 26 and 28), and in mature sieve elements (Fig. 63). Within the latter, the virions do not become enclosed in a membrane, as may occur with TMV, but, if abundantly represented, are spread throughout the lumen of the cell, sometimes forming a dense aggregation near the sieve plate (Esau et al., 1967). Particles of beet yellows virus have not been observed in immature sieve elements and in mature or immature tracheary elements, but the survey must be continued; one could at least expect viral infection of differentiating sieve elements.

With regard to the relations to the conducting tissues of viruses other than TMV and BYV, Price (1966) observed presumed virus particles in the form of flexuous rods in phloem cells of lime plants infected with tristeza virus. The author did not specify the type of these cells, but from the micrographs supplied they appear to be parenchyma cells located in the neighborhood of mature sieve elements. Gerola, Bassi, and Giussani (1966) illustrated an immature sieve element of *Brassica chinensis* with a body containing spherical osmiophilic particles presumed to be turnip yellow mosaic virions. Gerola, Bassi, Lovisolo, and Vidano (1966), studying wheat infected with maize rough dwarf virus, found numerous spherical particles typical of the virus scattered in a phloem cell with sparse contents. The cell appears to be a mature sieve element. Shikata and Maramorosch (1966a) found wound tumor virus particles in various parenchyma cells of the hosts, in tracheary elements in the infected root of *Melilotus officinalis*, and possibly in a sieve element in leaf-tumor tissue of *Trifolium incarnatum*, as is suggested by Figure 26 in the article. Chambers and Francki (1966) observed bacilliform particles of lettuce necrotic yellows virus in immature tracheary elements of *Nicotiana tabacum* infected with this virus. Numerous particles, indistinguishable from those of the isometric mottled crinkle virus of artichoke, were present in immature tracheary elements of the host, fewer in mature ones (Russo, Martelli, and Quacquarelli, 1967).

The information on cell types invaded by the different plant viruses is obviously too meager for a useful discussion of this aspect with reference to the selective or nonselective behavior of the virus in the host. A larger number of viruses must be examined, including such that are more strictly limited to the phloem than is the beet yellows virus. If the absence of beet yellows virions in the tracheary elements should prove to be a constant feature, it might be one of the expressions of the selective type of invasion of the host by this virus. The nonselective character of tobacco mosaic virus agrees with the occurrence of its particles in all kinds of cells in the plant body.

4

INTERCELLULAR TRANSPORT OF VIRUSES

Viruses are known to move in the host from one cell to another, slowly in parenchyma tissue, rapidly in the phloem or the xylem (cf. Esau, 1967b). The movement between parenchyma cells or between such cells and the sieve elements is commonly assumed to occur through plasmodesmata, the movement from sieve element to sieve element through sieve-plate pores. If virus movement in the xylem is rapid, it may be taking place in vessels in which the end walls of the individual vessel members are perforated at maturity so that a free passage between cells is realized. Movement through tracheids would require a transfer of virus particles through unperforated wall, possibly through the membranes of the pits. These membranes are readily passed by the xylem sap, but it is not known to what extent they are penetrable for virions. If, at maturity, the pit membranes were as loosely constructed as the primary wall parts in the tracheary elements shown in Figure 60, passage of virus particles through them would not be inconceivable.

The ultrastructural features of plasmodesmata suggest

that without some special modification these passageways between living cells may not be freely accessible to such large units as the complete virus particles. According to a considerable amount of data, plasmodesmata have a complex structure (Figs. 64–68). A membrane, the plasmalemma, lines the plasmodesmatal canal; a tubule (or a solid core) occurs in the center of the canal; a matrix (hyaloplasm) surrounds the tubular core. A further complexity is introduced by the usual association of endoplasmic-reticulum cisternae with the plasmodesmata (Figs. 64 and 65). In fact, the cisternae appear to be connected to the core. In successful sections the membrane of the endoplasmic-reticulum cisterna facing the plasmodesma is seen to be enlarged into a funnel-shaped protrusion which appears to be continuous with the tubule of the plasmodesma (Figs. 65 and 68; López-Sáez, Giménez-Martín, and Risueño, 1966, assume that both membranes of the endoplasmic-reticulum cisterna are invaginated at the plasmodesmatal opening). Plasmodesmata are often branched (Fig. 64). Between the sieve element and a parenchyma cell (which may or may not be a companion cell) the branching is one-sided: a single pore on the sieve-element side opens into several branches on the side of the parenchyma cell. Figure 69 illustrates such plasmodesmata in longitudinal section. One of the notable demonstrations in this view is that the plasmodesmata are not necessarily localized in the pit region. The plasmodesma to the right passes through a thick part of the wall to one side of the pit. The other peculiarity in evidence is that on the parenchyma side (above) the wall in the region of the plasmodesmatal branches is thickened. Developmentally, this thickening must have made possible a wider spacing of the branches in the inner layers (that is, those nearest the protoplast) of the wall. In sections cut through the wall on a tangent, the branched part of a plasmodesma may be recognized by the grouping of the cross sections of branches (Fig. 70). The single plasmodesmatal canal on the sieve-element side is slightly wider than those of the branches (Fig. 70, *arrows*) and is frequently surrounded by callose (Figs. 66, 69, and 70). Endoplasmic reticulum is associated with the branched plasmodesmata on both sides of the wall (not clearly shown in Fig. 69 but illustrated for the parenchyma-cell side in Fig. 68 and for the sieve-element side in Fig. 66).

The opinions regarding the detailed structure of plasmodesmata are not unanimous. Some authors favor the concept that the endoplasmic reticulum is continuous through the pore (e.g., López-Sáez et al., 1966); others consider that the core of the plasmodesma does not represent an endoplasmic-reticulum tubule and that it is only in contact but not continuous with the cisternae of the endoplasmic reticulum facing the plasmodesmatal openings (e.g., O'Brien and Thimann, 1967). Regardless of the exact structure of the plasmodesma, the covering of the plasmodesmatal openings with endoplasmic-reticulum cisternae and the partial obstruction of the plasmodesmatal canal by the core would seem to constitute substantial impediments to the passage of virus particles.

The sieve areas, which provide the communication channels between sieve elements, particularly those of the sieve plates (only these are considered in the following), appear to be more penetrable for particles of the size of virions than are plasmodesmata. Electron microscopy has shown that the formation of the sieve-plate pores involves removal of a considerable amount of wall material around the plasmodesmata initially present at the pore sites. The plasmodesmata and the associated endoplasmic-reticulum cisternae disappear, and a free communication between two superimposed sieve elements is established. As the tonoplast breaks down, this continuity pertains not only to the cytoplasm but also to the former vacuolar contents. A well-known phenomenon is the plugging of the pores with P-protein (slime) commonly seen in sieve elements of dicotyledons in material prepared for microscopy with the conventional method of cutting the fresh material before fixation. In Figure 71, P-protein with tubular components (cf. Fig. 72) is shown accumulated on one side of the sieve plate (slime plug) and filling the sieve-plate pores, which are also partly constricted by callose. The prevalent concept is that the plugging is a response to the release of pressure at the cut end of the sieve tube and the resulting rapid flow of cell contents toward the artificial sink. The less liquid material, the P-protein, passes through the sieve plate only partially and accumulates on the opposite side as a slime plug. This phenomenon is difficult to control (cf. Esau, 1966), but if plug formation is prevented, the sieve-plate pores appear open (Fig. 73).

The relation of virus particles to the channels of communication through cell walls has been successfully investigated for one virus thus far, the beet yellows virus (Esau et al., 1967). Particles of this virus were seen in bundles within plasmodesmata interconnecting two parenchyma cells (Fig. 74), within the branched plasmodesmata connecting sieve elements with the parenchyma cells (Figs. 75–78), and within the pores of sieve plates (Figs. 79 and 80). Both the longitudinal and the transectional views of plasmodesmata containing virus particles show no cores and no endoplasmic-reticulum associations. When and in what manner these structural features are removed in the presence of virus is an unanswered question. The virus particles are longitudinally oriented within the pores and flare out toward the cell lumen as though there were no obstacle to their penetration of plasmodesmata (Fig. 76).

Tobacco mosaic virions have not been seen in any pores within the walls of infected cells, either in the narrow canals of plasmodesmata or the wide openings in the sieve plates. The causes of the negative results of studies in this direction may have been that an insufficient amount of material was examined or that TMV passes through the walls in some other form than complete virus particle. Singh and Hildebrandt (1966), in studying movements of TMV inclusions in living infected callus cells, thought they saw passage of needle-shaped crystals through the walls and the entrapping of such needles in walls in formation during cytokinesis. These phenomena are yet to be investigated by means of the electron microscope. In some viruses the particles appear to be too large to pass through the plasmodesmata of their hosts. According to Kitajima (1965), the spherical particles of the Brazilian tomato spotted wilt are 80 to 120 mμ in diameter, whereas the plasmodesmata in its solanaceous hosts are 20 to 100 mμ wide. Lee (1967) concluded that the dimensions of the bacilliform particles of wheat striate mosaic virus would prevent their passage through plasmodesmata.

The arrangement of the beet yellows virus particles in the sieve-plate pores, parallel to one another, indicates a free access to these pores. The particles occupy the same kind of position in the pores as normally does the P-protein in material processed without special precautions. In Figure 80, in fact, one pore is seen filled with virus particles, the other (shown only in part) with P-protein (compare with Fig. 71). Some P-protein in Figure 80 was directed also toward the virus-containing pore (*arrow*), and the pore to the extreme left in Figure 79 contains both virus and P-protein. It seems obvious that the P-protein and the virus can substitute for one another in the pores, that both occupy the same space in the lumen of the cell, and that both are subject to displacement by the release of pressure when the sieve element is cut. These relations seem significant with regard to the state of the sieve-plate pores in the intact sieve tube: the pores are not plugged with P-protein in the undisturbed plant; if they were plugged, they would be inaccessible to the virus particles.

Physiological studies, especially those of Bennett (1940, 1956), have convincingly shown that virus translocation in the phloem is subject to the same laws as is the movement of organic food materials. The concept has been advanced therefore that viruses move with the translocated food. Moreover, the observation that in this movement virus particles in passing through the plant can leave some portions of it noninfected was interpreted as indicating that the virus is transported passively as a particulate matter. Electron microscopy shows that in plants infected with beet yellows virus this particulate matter may be the complete virus particle. The manner of invasion of sieve-plate pores by virus particles in cut material agrees with the concept that these pores are normally open and permit a mass flow of contents. If virus particles are transported as such in the translocation stream within the sieve tubes and if under release of pressure they behave as does the P-protein in plugging the pores, one should raise the question as to whether the P-protein also is transported from sieve element to sieve element during the movement of photosynthates. Some of the P-protein is probably carried along with the food but its major role in functioning sieve elements may be one to provide a submicroscopic skeletal framework to support the highly labile contents of the cell. Upon injury, the framework could become transformed into the system of P-protein strands visible with the light microscope and interpreted as normal structural elements of the intact cell by Evert and Derr (1964).

As was discussed in Chapter 3, one can postulate that the virus transported in the sieve elements has two sources. This virus may be the progeny of the virions that invaded the young sieve element or its precursor and underwent a replication in this cell. Figures 57–59 show for TMV that a virus may be present in differentiating sieve elements, although this phenomenon has not yet been demonstrated for the beet yellows virus. The second source of the translocated virus would be the parenchyma cells that are contiguous with the sieve elements. Such cells are frequently almost filled with virus particles and in BYV infections the same kind of particles occur in plasmodesmata leading to the sieve elements. The movement between the parenchyma cells and the sieve elements is, undoubtedly, a two-way exchange. Virus can be assumed to be carried by the water associated with the photosynthate delivery from the parenchyma cell to the conduit. Parenchyma cells also withdraw sugars from the sieve elements and the movement in that direction would provide the mechanism for carrying the virus from the conduit to the parenchyma cell. If this cell is still free of virus, a new infection would be established. Since the phloem is spread as a continuous system through the entire plant, and since the food moves toward sinks in various parts of the plant, particularly the growing parts, the virus transported with the food can effectively invade the entire plant and continue to invade the new growth increments.

A discussion of the mechanics of virus movement in the water conduit is burdened with more uncertainties than is that concerned with translocation in the phloem. In fact, natural translocation of a virus in the xylem has been unequivocally demonstrated for a single virus, the southern bean mosaic virus (Schneider and Worley, 1959a, b; cf. also Schneider, 1965; Esau, 1967b). The especially problematic aspect of virus movement in the xylem is the interchange of virus between living parenchyma cells and the nonliving conduit. The movement from the tracheary cell into a living parenchyma cell may involve some form of pinocytosis, but it is questionable whether this method of entry would be available for the movement in the opposite direction, that is, from a living cell into one devoid of a protoplast, unless pinocytosis can become inverted (Holter, 1961) and the virus secreted into the xylem sap. Schneider (1965) suggests that virus particles may reach the conducting tracheary elements by invading first the immature elements, then progressing through partially differentiated cells, and finally entering the mature conduits through degenerating membranes. It is also conceivable that virus particles invading an immature cell become mobile after that cell matures.

The spread of virus in a tissue may occur through its distribution among the products of cell division as has been clearly demonstrated in TMV-infected tissue cultures (Chandra and Hildebrandt, 1966, 1967). Electron microscopy shows that tobacco mosaic virus is well represented in young cells engaged in meristematic activity (Figs. 81 and 82), and the views of infected cells in mitosis and cytokinesis suggest that the virus is randomly distributed between the daughter cells, as are the host-cell organelles. The sequence of events that brings about the division of a cell by a new wall is becoming well understood at the ultrastructural level (cf. Cronshaw and Esau, 1968a; Hepler and Newcomb, 1967). Vesicles derived from dictyosomes move toward the equatorial region of the cell, possibly under the influence of some directional activity of the microtubules of the phragmoplast. The vesicles fuse into the cell plate (Fig. 83) which at first is limited to the region originally occupied by the metaphase plate. The activity of the phragmoplast progresses centrifugally, extending the cell plate until the latter reaches the mother-cell wall.

No disturbances in cytokinesis or the earlier mitosis were perceptible in young TMV-infected cells, even in the presence of large amounts of virus. The details of these phenomena were entirely similar to those previously recorded for noninfected leaves of *Nicotiana tabacum* (Cronshaw and Esau, 1968a). This observation agrees with the experience of Chandra and Hildebrandt (1966, 1967) who found that single TMV-infected cells are capable of division and formation of colonies and even of whole plants. In the dividing cells, as in the quiescent ones, the virus is distributed in various ways. The aggregates of particles may occur in cell portions removed from the region where the cell plate is formed eventually (Fig. 83). Such aggregates apparently are not affected by the division; they become incorporated into the new cells in accordance with their position before the division (because of cyclosis this position may be only temporary, of course), and it is easy to visualize that one of the two daughter cells could get all the virus that was present in the mother cell and the other remain noninfected. According to Chandra and Hildebrandt (1966), in tissue cultures derived from TMV-infected cells only about 40 per cent of cells contain virus. A virus aggregate, however, may lie in the equatorial plane (Fig. 82) or actually in the path of the forming cell plate (Fig. 84). No indications were found that virus in such positions interferes with the progress of cell-plate formation. In Figure 85 the virus aggregate appears disturbed in front of the cell plate. This view suggests that an aggregate may be mechanically ruptured if it is in the way of the cell plate. Figure 86 could be taken as an illustration of the end result of such rupture: the two aggregates may have been derived from a single aggregate which was divided between the two cells in the cell-plate region. Possibly, also a virus aggregate may be moved away from the cell-plate region by cytoplasmic movement.

An irregularity was observed in cell-plate formation where an X-body was located close to the plane of division. In Figure 87 the final stage of normal cell-plate formation is characterized by the presence of phragmoplast microtubules close to the mother-cell wall (W) and of fusing cell-plate vesicles, which were about to make contact with this wall. As is usual, the phragmoplast microtubules in this figure extend to both sides of the growing cell plate. In Figure 88, cell-plate formation was also nearing its end but here phragmoplast microtubules occurred only on one side of the cell plate. On the other side, the X-material lying next to the plate seems to have prevented the formation of part of the phragmoplast.

Although cell division appeared to progress normally in most of the material described above, some viral strains may have a greater effect on the process than others. Gol'din, Agoeva, and Tumanova (1967), studying living cells from tobacco tissue cultures infected with the Kasakh strain on TMV, reported that fusiform, needle-shaped, or fibrous looped inclusions were seen interfering with the completion of cell-plate formation or forcing the plate to assume an inclined position. Sometimes, however, the inclusion was broken by the growing cell plate and was thus divided between the two daughter cells. The authors also envisage suppression of cell division by viral inclusions penetrating the nucleus.

The material reviewed in this chapter indicates that viruses may invade new cells as complete particles either by passing through pores in the cell wall or by being distributed between daughter cells during cell division. Since, however, this evidence is based on studies of two viruses only, other modes of virus spread from cell to cell must be visualized. Tobacco mosaic virus particles have not been seen in wall pores as yet. Although further search may reveal such relation, it is conceivable that some viruses move in the form of naked infective nucleic acid, the demonstration of which within the host cells has not been achieved as yet.

5

RESPONSES OF INFECTED CELLS
TO PRESENCE OF VIRUS

IN considering the effects of viruses on the host cells one must differentiate between the activities that lead to the formation of virus and the phenomena that cause degeneration of cell components and frequently death of cells. Weintraub and Ragetli (1964a) see an evidence for the existence of the two viral functions in the comparative effects of TMV on the leaves of *Nicotiana glutinosa*, in which infection results in the development of localized lesions, and on those of *N. tabacum*, which become systemically infected. The leaves of *N. tabacum* had about 10^5 times as many virus particles per unit of fractionated homogenate scanned under the microscope as did those of *N. glutinosa*. At the same time, the locally infected tissues of *N. glutinosa* showed an abnormal rise in respiration, a reaction not observed in infected *N. tabacum*. The rise in respiration in *N. glutinosa* was traced to an increase in the number of mitochondria and the resulting rise in enzyme (succinoxidase) activity (Weintraub, Ragetli, and Dwurazna, 1964). Weintraub and Ragetli (1964a) concluded that the destructive effects of the infection in *N. glutinosa* were caused by a metabolic change not directly related to viral multiplication.

Diener (1963) expressed the thought that if the pathologic alterations in the host plant did not result solely from the production of viral components, the invading nucleic acid molecule must contain information for the initiation of the derangements. In other words, the viral RNA must possess independent genes, some of which are involved in its own replication and the production of virus-specific proteins; others cause changes in the metabolism of the host plant. Ragetli (1967) presented a similar concept when he stated that viral replication may be associated with excessive genetic information directing protein synthesis by the host and affecting thereby the metabolism of the latter. Thus, pathogenesis in virus-diseased plants results not from introduction of injurious metabolites as it does in microbial infections, but from addition of genetic determinants to the host cell. As Diener (1963) characterized it, pathogenesis in viral infections is not a metabolic but a genetic phenomenon.

The degenerative changes in virus-infected plants assume various aspects as seen with the light microscope. There may be a depression of development (hypoplasia) or a stimulation of growth of cells in size (hypertrophy) or in number (hyperplasia). Necrosis is a frequent symptom which may occur in association with growth abnormalities or without such. The investigations in this area have been reviewed consecutively (cf. Esau, 1967*b*). The information on the ultrastructural features of the degenerative changes is still meager. Moreover, the descriptions of the changes are often accompanied by illustrations in which the recorded abnormalities appear to have been caused, at least in part, by technical deficiencies. Possibly virus-infected cells are more sensitive to technical manipulations than are healthy cells so that the degenerative changes become artificially exaggerated in the processed material.

DEGENERATIVE CHANGES IN THE CHLOROPLASTS

Destructive effects of viruses as seen with the electron microscope have been mentioned most frequently with regard to the chloroplasts. In leaves of Chinese cabbage infected with turnip yellow mosaic virus, Gerola, Bassi, and Giussani (1966) observed considerable disturbances in the internal organization of the chloroplasts from the yellow and whitish leaf areas. The stroma lamellae were fragmented and the grana thylakoids swollen. Vesicles were formed by invagination of the inner membrane of the envelope. In most severe form of degeneration the chloroplasts showed extremely reduced thylakoid systems and were deeply lobed. Chalcroft and Matthews (1966), studying the same virus in the same host, recognized degenerative changes only in the chloroplasts and concluded that these changes varied in relation to the type of symptom in the leaf, namely, whether the symptom was a change in color in a systemically infected leaf, or a lesion in a leaf with localized infection. Some of the changes in the chloroplasts were more than normal or less than normal starch accumulation, scarcity or absence of grana, presence of phytoferritin, general swelling of the plastid. Shalla (1964) and de Zoeten (1966) described ameboid form changes in the chloroplasts in infections with TMV and tobacco rattle virus, respectively. According to Gardner (1967), the chloroplasts in infections with barley stripe mosaic virus contained few grana, the intergrana lamellae dominating the picture, and there were some paired membranes in distorted configurations. Arnott and Smith (1967) reported reduction in grana thylakoids in sunflower infected with an unidentified virus. Gerola and Bassi (1966), studying *Zea* infected with maize rough dwarf virus, found the grana replaced by long lamellae in parallel arrangement. Vesicles were present in the stroma along the chloroplast envelope. Dispersal of grana or of thylakoids is sometimes mentioned in connection with plastid disintegration (e.g., Israel and Ross, 1967). Weintraub and Ragetli (1964b) stated that in lesions induced by TMV in leaves of *Nicotiana glutinosa* the starch grains increased in size and, later, the chloroplast membranes became ruptured. Ragetli (1967) postulated that chloroplast degeneration was caused by the activation of the normally structure-bound acid phosphatase.

Some of the degenerative changes in the chloroplasts mentioned above, especially the modification of the organization of the thylakoid system, were observed in beet leaves infected with the beet yellows virus. The sugar beet chloroplasts show the typical structure of these organelles in dicotyledons (Figs. 89–91). Their double-membraned envelope encloses a relatively dense stroma in which a system of lamellae (thylakoids) is differentiated into grana and intergrana lamellae. Sometimes the thylakoid system is displaced toward the side away from the cell wall. This phenomenon is not limited to healthy material and may be observed after both aldehyde- and permanganate-based fixations (Figs. 89 and 91); it may be an artifact. Among the abnormalities noted in the thylakoid system of chloroplasts in leaves infected with beet yellows virus were a partial breakup of the system into apparently discrete units (Fig. 92), disorientation of grana (Fig. 93), poor differentiation of grana or their substitution by stacks of long thylakoids (Figs. 94 and 95). Some plastids contained rather large vesicles in the peripheral part of the stroma (Fig. 94), which probably originated as invaginations of the inner membrane of the envelope (Fig. 95). Such vesicles occurred also in chloroplasts in noninfected plants (Fig. 90), but they were more conspicuous in the infected material. The vesicles apparently can become extruded from the chloroplast (Fig. 94). Occasional plastids in the infected leaves contained opaque inclusions of unknown identity (Figs. 94 and 95, *arrows*). Some abnormal plastids had a denser stroma than usual (Figs. 94 and 95), and in advanced stages of degeneration the internal structure was largely replaced by an amorphous matrix (Fig. 96). No rupture of envelope and release of grana were observed.

Some viral infections are associated with an increased accumulation of osmiophilic lipid globules. Globules normally present in plastids may increase in size or in number or both. This symptom was reported, for example, for the unnamed viral infections of *Amaranthus lividus* (Rubio-Huertos and Vela-Cornejo, 1966) and of the sunflower (Arnott and Smith, 1967), and for the localized lesions in TMV-infected leaves of *Datura stramonium* (Carroll, 1966). Nakata and Hildebrandt (1967) mentioned the presence of numerous lipid bodies in the plastids in callus derived from leaf cells of tobacco infected with tobacco mosaic virus. Israel and Ross (1967) observed lipid globules in plastids located in the zone of surviving cells bordering the collapsed cells of localized lesions induced by TMV in *Nicotiana tabacum* var. Samsun.

An excessive accumulation of osmiophilic lipid material in chloroplasts is prominent in sugar-beet leaves infected with the beet yellows virus. Osmiophilic globules, or plastoglobuli (Lichtenthaler, 1966), are common components of healthy sugar-beet chloroplasts. Aldehyde and osmium tetroxide fixatives usually preserve them as spheres (Figs. 89 and 90), potassium permanganate as irregular bodies (Fig. 91; star bodies of Strugger, 1957). The globules may be uniformly osmiophilic (Fig. 89), or they may be less osmiophilic in the center than at the periphery (Fig. 90). In leaves infected with BYV the plastoglobuli increase in size so that they often cause a displacement of grana and may even deform the plastid (Figs. 97 and 98). Starch grains also occur in these chloroplasts (Fig. 98), whereas normally starch is absent in sugar beet leaves. Plastids containing globules and starch were found occasionally as twin structures, with two separate thylakoid systems enclosed in the same stroma and bound by one envelope (Fig. 99). In some cells, osmiophilic globules were present in the cytoplasm (Fig. 100).

In some samples of infected leaves the lipid plastoglobuli in the chloroplasts had striking surface ornamentations in the form of paracrystalline systems of rods (or possibly tubules; Fig. 101). Cronshaw, Hoefert, and Esau (1966) previously observed this structure and speculated that the rods may be virus particles accumulated on the surface of the lipid globules. Further examination of the phenomenon did not support the identification of the rods with the virus. The rods on the lipid globules are shorter than the typical BYV particles, and the chloroplasts appear to contain no free virus particles. When the staining of the sections on the grids with uranyl acetate and lead citrate is omitted, the virus particles in the cytoplasm are barely discernible, whereas the rods on the globules are as strongly electron opaque as they are in the stained sections. (Their dark appearance in this study was an indication of osmiophily.) Another important feature is that the rod complex appears to be an integral part of the globules rather than a mere surface accumulation. In different globules the rod systems are of different depths, a feature giving the impression that the peripheral part of the globule gradually assumes a paracrystalline state which, with time, affects more and more of the substance of the globule

(Figs. 102 and 103). The degree to which the amorphous core of the globule may be deformed by the development of the rods is strikingly illustrated in Figure 103. (The spreading of the rods away from the globule in this figure may have resulted from some extraneous damage.) The transformation into the rod system sometimes affects only localized regions of the globules (Fig. 104). Some globules were found to have a different kind of surface elaboration: a layer of material of alveolate or foamy appearance, clearly delimited from the core and having a smooth surface (Fig. 105). The alveolate layer was of different depth in different globules and was partly extruded from the chloroplast in some views. It is not certain whether any artifacts were involved in the formation of the alveolate layer or its occasional extrusion from the chloroplasts.

A search in the literature revealed one reference describing a patterned structure of osmiophilic globules. Silaeva and Shiryaev (1966), in studying the effect of various fixatives on the osmiophilic globules, found that a mixture of osmium tetroxide and potassium permanganate produced a peculiar differentiation in the globules. A lightly osmiophilic inner part was surrounded by a thick paracrystalline layer composed of spherical particles and thin threads interconnecting the particles. The sharp delimitation of the peripheral layer from the core and along the surface shown in the illustrations accompanying the article suggests a similarity of structure to the globules with the alveolate surface layer as depicted in Figure 105.

The accumulation of osmiophilic material in chloroplasts is not a specific response to viral infection. Many other factors may induce the increased development of plastoglobuli. As observed in *Vicia faba*, the plastoglobuli become larger and apparently also more numerous when the leaf ages (Greenwood, Leech, and Williams, 1963).

Butler (1967) noted accumulation of large osmiophilic globules in chloroplasts of senescing cotyledons in *Cucumis sativus*. Globules increase in number in etiolated plants (Lichtenthaler, 1966) and are abundant in chlorotic plants (Strunk and Wartenberg, 1960). Their development is promoted by phosphorus deficiency (Thomson, Weier, and Drever, 1964). An increase in osmiophilic matter occurs in chloroplasts that differentiate into chromoplasts (e.g., Thomson, 1966), a phenomenon that may be reversible (Thomson, Lewis, and Coggins, 1967). Usually the increased development of globules is associated with degradation of grana, and therefore the material of the globules has been interpreted as lipid derived from grana membranes (e.g., Strunk and Wartenberg, 1960). It further has been proposed that the lipid of the plastoglobuli is utilized in the formation of grana during plastid ontogeny, but Greenwood, Leech, and Williams (1963) found no support for this view. In *Vicia faba* chloroplasts, globules continued to enlarge during active lamellar growth. Lichtenthaler (1966) also observed no relation between the plastoglobuli and the development of grana, for globules were present in the chloroplasts in all stages of their development. Greenwood, Leech, and Williams (1963) suggested that the globules may be a general deposit of insoluble lipid material just as starch is a deposit of insoluble carbohydrate. According to this view, the globules may be expected to contain a wide range of lipids, not necessarily the same in different plants, nor at different times in the same plant. The variation in the degree of osmiophily found among the globules (e.g., Figs. 89, 90, 92, 94, 95) may be ascribed to variation in the chemistry of lipids stored. Globules contain lipophilic quinones (Bailey and Whyborn, 1963) and, in the chromoplasts, may become the sites of accumulation of carotenoids and other lipid-soluble pigments (Greenwood et al., 1963). The plastoglobuli do not coalesce when they are extracted from chloroplasts but no bounding membrane was observed (Greenwood et al., 1963). As was mentioned before, osmiophilic globules may occur in the cytoplasm and in this position also they may show different intensities in the osmiophily (Figs. 100 and 106). Lipid material is known to accumulate in the cytoplasm of normal cells. O'Brien (1967) observed lipid droplets in the epidermal cells of the coleoptile of *Avena* and reviewed the concept that the material may be the cutin precursor. Peveling and Lichtenthaler (1967) found cytoplasmic lipid inclusions in cells of onion scales, particularly abundantly in the laticifers (not recognized as such by the authors), and interpreted the material as a storage product of the starch-free onion bulbs. O'Brien (1967), as well as Peveling and Lichtenthaler (1967), suggested that the cytoplasmic globules were the entities that some authors described as spherosomes on the basis of light-microscope studies. According to Butler (1967), the cytoplasmic osmiophilic globules in senescing cotyledons of *Cucumis* were derived from ruptured chloroplasts. In cells of the sugar beet infected with beet yellows the cytoplasmic lipid globules appeared to be independent of those in the chloroplasts. In Figure 106 the globules are dispersed within the virus-containing cytoplasm with no obvious association with the abnormal but intact plastid.

Plastoglobuli are evidently ubiquitous components of plant-cell protoplasts, especially of the plastids, and their development is influenced by a variety of external and internal environmental conditions. The excessive development of lipid globules in plants infected with some viruses may safely be attributed to metabolic changes that increase the deposition of insoluble lipid materials. Increased deposition of starch in the chloroplasts, frequently mentioned in connection with viral infections, may be a concomitant phenomenon (Figs. 98, 99, and 106).

Chloroplasts of beet leaves infected with the beet yellows virus often contain finely granular material, with the granules mostly in orderly arrays (Fig. 107). Cronshaw, Hoefert, and Esau (1966) depicted the same material and interpreted it as consisting of rods. Moreover, the authors identified these presumed rods with those appearing on the surface of the osmiophilic globules. Actually, the aggregates not associated with the globules consist not of rods but of granules. When these granules are in crystalline arrays certain sections of the rows of granules give the impression of rods (Figs. 107 and 108), but when they are arranged at random no rods can be recognized (Fig. 109).

Particles similar to those just described have been illustrated by several investigators and interpreted as phytoferritin (Catesson, 1966; Hyde et al., 1963; Newcomb, 1967; Robards and Humpherson, 1967; Sitte, 1965, p. 142; Sprey, 1965). The particles depicted, but not identified, in papers by Sitte (1961) and Schnepf (1961) are regarded also as phytoferritin by Hyde and co-workers (1963). Ferritin was recognized only in animal cells until Hyde, Hodge, and Birnstiel (1962) reported the discovery of this material in ribosome preparations from pea embryos. The authors named the material phytoferritin. According to Hyde, Hodge, Kahn, and Birnstiel (1963), the phytoferritin is an iron-protein complex. In negatively stained extracts of pea embryos the particles were found to be doughnut-shaped and to have a dark center. Particles corresponding to the dark centers appeared in unstained purified preparations of phytoferritin as dense, square or polygonal granules. Hyde and co-workers (1963) interpreted the dense particles as the iron-rich cores of the larger particles seen in the negatively stained preparations. The core is embedded in protein which makes the particle appear doughnut-shaped. The core is not homogeneous but consists of several subunits, most commonly four. Their arrangement around a central point imparts a square shape to the particle (Fig. 109). Robards and Humpherson (1967) have called attention to the circumstance that phytoferritin particles (actually the iron-rich cores of the particles) are clearly visible in preparations that are not stained with uranyl acetate or lead citrate. The presence of iron makes the particles opaque to the electrons. Figure 108 shows a portion of a chloroplast from an unstained preparation of a BYV-infected beet leaf: the particles appear as dense as do those in the stained preparation (Fig. 107). One may conclude that the squarish granules are not related to the virus but are most likely phytoferritin.

The suggestion made by Cronshaw, Hoefert, and Esau (1966) that the rods appearing on the surface of the osmiophilic globules and the phytoferritin arrays might be viral material is the second instance that normal cell components are suspected of being the beet yellows virus. Earlier, Engelbrecht and Esau (1963) thought that protein crystals which they observed in chloroplasts of beet leaves infected with two different strains of beet yellows virus could be the virus. Since then, crystals of the same type were seen in noninfected material (Cronshaw et al., 1966).

The significance of the accumulation of phytoferritin in beet-leaf chloroplasts infected with BYV can only be surmised. Hyde and co-workers (1963) advanced the concept that phytoferritin serves for the formation of iron-containing components of the photosynthetic apparatus (e.g., ferredoxin). Robards and Humpherson (1967) find this concept acceptable and suggest that the deposition of phytoferritin may be the safe means of storing an iron-containing protein in a nontoxic form when the iron is not utilized in photosynthesis. This suggestion agrees with the occurrence of phytoferritin in cells in early stages of differentiation or in those that for some reason are incapable of performing full photosynthetic activity. Accumulation of phytoferritin in sugar-beet leaves infected with BYV could be an expression of impaired photosynthetic activity.

The chloroplasts characterized by accumulations of lipids and starch were seen in many cells that, in the same sections, revealed no virus particles. In contrast, the chloroplasts with more severe abnormalities and showing evidence of approaching necrosis occurred in cells with large amounts of virus. The difference suggests that the degenerative changes have complex causes and do not follow a single sequence. In some, the principal change is the accumulation of insoluble lipids and carbohydrates, in others the membrane system is disorganized; and there is no evidence that one type of change necessarily follows or precedes the other.

The effect of tobacco mosaic virus on the chloroplasts of infected hosts has not been adequately investigated with the electron microscope. Light-microscope studies of systemically infected plants of tobacco indicate that in the yellow-green areas of mosaic leaves the chloroplasts are smaller in size and less numerous than they are in the dark-green areas of the same leaves, and that they are deficient in chlorophyll (cf. Esau, 1948). As to the differentiation of the mesophyll, the yellow-green areas are hypoplastic. The relation of the abnormalities in the yellow-green areas to the presence of virus is indicated by the observations that the yellow-green mesophyll contains considerably more virus than does the dark-green mesophyll, both with regard to infectivity tests and to amount of viral inclusions as seen with the electron microscope (Atkinson and Matthews, 1967). The degeneration of plastids is particularly severe in the localized lesions (e.g., Israel and Ross, 1967). In the material studied in connection with the present report, one of the most frequent symptoms of infection with TMV in the chloroplasts was the inclusion of virus particles in the plastid stroma (Fig. 42). Structurally, these chloroplasts usually appeared to be normal. Vesicles derived from the inner layers of plastid envelope were seen sometimes in chloroplasts of infected cells, but such vesicles occurred also in some chloroplasts of noninfected plants (Fig. 110). A modification of the thylakoid system was observed in chloroplasts of older mosaic leaves. Figure 110 illustrates the structure of a normal chloroplast from a noninfected plant of *Nicotiana tabacum*, with its characteristic grana and intergrana thylakoids. In the chloroplast from the infected leaf (Fig. 111) grana development was largely suppressed so that lamellae with long profiles predominated. They formed a reticulate pattern at the margins of the plastids. Osmiophilic globules were somewhat more numerous and larger in the degenerated plastid than in the normal. Two other modifications in the structure of chloroplasts encountered in infected leaves are shown in Figures 112 and 113. Two sets of apparently normal thylakoid systems are combined into a twin structure within one envelope in Figure 112. Arnott and Smith (1967), studying a mosaic disease in the sunflower, observed a twinning of chloroplasts of somewhat different nature. The two plastids had individual envelopes but were associated with one another as though they had failed to separate after a longitudinal division. The chloroplast in the TMV-infected cell of tobacco in Figure 113 shows a reticulate system of membranes in the grana-free region facing the cell wall. The significance of the modifications shown in Figures 112 and 113 is obscure.

DEGENERATIVE CHANGES IN SIEVE-ELEMENT PLASTIDS

Previously, the concept was reviewed that viruses are transported in the sieve element passively with the organic food. This concept, however, should not be taken to mean that the sieve element is not affected by the presence of virus in the phloem. Light-microscope investigations have shown that immature and recently matured sieve elements may become necrosed in virus-infected plants (e.g., Esau, 1958), and the electron microscope has revealed that sieve-element plastids may undergo a conspicuous disorganization (Hoefert and Esau, 1967). Whether such degeneration of plastids depends on the presence of virus in the sieve-element specifically and whether, if it does so, the virus must enter the sieve element while it is still differentiating, needs to be determined.

Degeneration of sieve-element plastids was reported by Hoefert and Esau (1967) for sugar beets infected with the curly top virus. There were changes in the membrane system, a displacement and reduction in size of the fibrous proteinaceous ring characteristic of these plastids in the sugar beet, and an accumulation of osmiophilic material. The virions of curly top have not been seen as yet in situ in infected cells, and thus their presence or absence in cells with degenerated plastids was not ascertained.

Infection with the beet yellows virus also may induce degeneration of sieve-element plastids, the details of which resemble those observed in curly top virus infections. Figures 114–120 compare normal sieve-element plastids (Figs. 114–116) with those showing abnormalities (Figs. 117–120). The sieve-element plastid in the sugar beet (and other Chenopodiaceae and some Aizoaceae) has certain features distinguishing it from sieve-element plastids in the majority of other investigated dicotyledons. First, the plastid contains a proteinaceous inclusion (cf. Esau, 1965) in the form of a fibrous ring (Figs. 114 and 115). Second, in the leaves the plastids do not accumulate starch (they may do so in the root) as is typical of sieve-element plastids in many other dicotyledons. Third, the surface of the plastid is covered with tubules which appear to be those of the P-protein. The identity of the two kinds of tubules may be surmised from their similarity in morphology (Fig. 115) and from the timing in the establishment of the association with the plastid. Immature plastids are devoid of tubules (Fig. 116) probably because they are located in cells in which the P-protein is still concentrated in the P-protein bodies (slime bodies). The fibrous ring surrounds a region containing membranous entities resembling sacculi in form and not organized according to an orderly pattern. Osmiophilic globules may be present among these sacculi (Esau, 1965).

In the degenerating sieve-element plastid the most conspicuous change affects the membranes. The sacculi lose their inflated appearance to a large extent, elongate considerably (Figs. 117 and 118), and may become compacted into granalike aggregates (Fig. 117). Lipid material becomes more conspicuous than in normal plastids (Fig. 118) and may form very large globules (Fig. 117). The fibrous ring is thinner than normal (Fig. 117). Without developmental studies it is impossible to judge whether plastids, as in Figures 117 and 118, developed abnormally or whether they changed after a normal development. It is also uncertain whether these plastids illustrate intermediate stages of degeneration, possibly preceding the more severe disorganization shown in Figures 119 and

120. In these figures the plastids are devoid of a matrix, the membranes are sparse, and, in the plastids in Figure 120, the small amount of fibrous material is in a dispersed state or is absent. The P-protein on the surface of the plastid envelope seems to remain unaffected in the degenerating plastids (Figs. 117–120).

The abnormal plastids just described occurred in sieve elements in which beet yellows virions were not detected. In fact, sieve elements with large accumulations of virus particles contained normal-appearing plastids. In Figure 63, for example, the plastid shows no abnormalities although it is surrounded by virions. The almost empty body next to the plastid, lacking a P-protein coating, is probably derived from the plasmalemma as an artifact. Similar structures occur in healthy phloem (cf. Esau, 1967a).

In tobacco infected with tobacco mosaic virus no obvious degenerative changes were recorded in sieve-element plastids, but as was observed with regard to the chloroplasts, apparently normal sieve-element plastids contained virus particles (Fig. 49).

PHLOEM DEGENERATION

In many viral infections degenerative changes result in necrosis of cells (cf. Esau, 1948). In phloem-related viruses such degeneration is initiated after the sieve element matures at a given level of shoot or root, in cells contiguous with the sieve elements. This relationship is commonly interpreted to mean that the degeneration is induced by the virus translocated in the sieve tubes and entering the noninfected cells from the conduit. With the light microscope the first evidence of degeneration in such cells is a dense staining of the protoplasts. Electron microscopy provides the means of investigating the structural details associated with this chromatic condition and of determining, specifically, whether virus particles occur in the affected cells. The phloem of root tips and of minor veins in the leaves is particularly convenient for the study of the relation between sieve elements and virus-induced cell degeneration. This phloem occurs at the terminals of the conducting system where the degenerative changes may be observed as they are initiated in relation to the maturation of the first sieve elements in those regions. The phloem of minor veins has the added advantage that the sieve elements are narrow and sparse, whereas the cells to be studied are comparatively wide and their topographic relation to the sieve elements is very clear (Esau, 1967a).

In the following paragraphs, the degenerative changes in the phloem, as seen with the electron microscope, are described with reference to the sugar beet infected with the beet yellows virus. For orientation, Figure 121 illustrates a portion of the phloem from a minor vein in a young healthy leaf. The two sieve elements in view are in contact with parenchyma cells on all sides. These parenchyma cells vary in degree of vacuolation, but show no obvious distinction between companion cells and phloem-parenchyma cells (cf. Esau, 1967a). Each of the two sieve elements in view is in contact with four cells, with three of which it is ontogenetically related (cells numbered *1-3*). In all three of these cells of a group the parietal cytoplasm is equally dense, and the cells probably play one and the same role in the exchange of materials between the sieve element and parenchyma cells, including the delivery of photosynthates to the conduit. The conspicuous organelles in the cells associated with the sieve elements are nuclei, chloroplasts, and mitochondria. The density of the cytoplasm is clearly related to the presence of numerous ribosomes, a feature further empasized in Figure 122, in which portions of the two parenchyma cells left of and below the sieve element show the numerous ribosomes at a high magnification. The sieve element in Figure 122 is not yet mature—its vacuole is still delimited by a tonoplast—but its ribosomes are already sparse and the endoplasmic reticulum has assumed the parietal position characteristic of mature sieve elements. No P-protein is evident in the cytoplasm except for a few tubules next to the plastid (*arrow*). The P-protein was probably still largely confined to slime bodies as is typical of young sieve elements. Incidentally, Figure 122 offers a comparison with regard to width between P-protein tubules (*arrow*) and the microtubules (*MT*) commonly present next to the plasmalemma in cells with growing walls. The plasmalemma appears as a double line along the wall in all three cells in view.

In Figure 123, illustrating a transection of phloem in a minor vein of a sugar-beet leaf infected with beet yellows virus, the three sieve elements in view are associated with parenchymatic cells. Some of these have extremely dense protoplasts; others have less dense protoplasts that contain large vacuoles. Selected cells in this figure are designated by numbers *1–6*. Cell *1* is tightly packed with virions; no cytoplasmic components were discernible in higher magnification micrographs of the same cell. It is probably a degenerated sieve element. Cell *2* shows mainly virus particles and peculiar vesicles, although a few mitochondria are in view also. (The identification of virus particles in the various cells in this figure was made by reference to higher-power electron micrographs of the same cells.) Cells *3* and *4* contain extremely dense lobed nuclei, large numbers of virions, and vesicles and vacuoles of various sizes. Cell *5* shows several transections of protrusions that had invaded the vacuole. Virus particles occur in the cytoplasm including that in the protrusions. At higher magnification ribosomes were discernible in cells *1–5*. Despite its contact with a sieve element, cell 6 shows no virus particles and no structural abnormalities except a large amount of starch in the chloroplasts.

If the virus particles present in the four parenchyma cells 2–5 in Figure 123 were derived from the contiguous sieve elements, the apparent absence of virions in the sieve elements (except in the presumed sieve element numbered 1) seems peculiar. Perhaps only a small number of virions need to be transported in the sieve tube to establish infections in adjacent parenchyma cells. There are views, on the other hand, in which both the sieve element and the contiguous parenchyma cells show virus particles in profusion (Fig. 124). It is conceivable that the large virus accumulations in sieve elements result from an early invasion of the cell and a multiplication of the virus in the still nucleate protoplast.

Figure 124 reveals that the density of the protoplast in a degenerating cell proximal to a sieve element in an infected leaf is accountable, at least in part, by the small volume of the vacuome and the packing of the cytoplasm with virions. As was discussed in Chapter 2, the multiplication of viruses in cells seems to force the cytoplasm to encroach upon the vacuole either by increase of the width of the parietal layer or by formation of virus-containing protrusions. Figure 106 graphically illustrates the filling of a vacuole, to the right, by such protrusions. Ribosomes, which make the parietal cytoplasm appear dense in normal cells (Figs. 121 and 122), seem to have been retained in the cell shown in Figure 124. Islands of ribosome-rich cytoplasm are scattered among masses of virions. Mitochondria also occur in these islands. The single chloroplast in view contains starch but otherwise shows no abnormalities. The nucleus in the lower portion of the cell is partly degenerated and contains a large aggregate of virions in the center. Figure 124 also depicts a common phenomenon in virus-infected tissues: virus and virus-induced symptoms are irregularly distributed among cells of the same region. The parenchyma cells to the left in the picture contain no virions and exhibit no pathologic changes except accumulation of starch in the chloroplasts.

The vesicles aggregated in the center of the virion-containing parenchyma cell in Figure 124 are the most outstanding feature observed in beet yellows-diseased cells. The higher magnification micrograph in Figure 125 shows that the vesicles occur in groups and are interspersed with aggregates of virions. Each vesicle has a densely stained membrane and contains fine fibrils, frequently joined in the center where a dot may be visible. Sometimes the vesicles of a group appear to have a common membrane (Fig. 95). The vesicles have not been seen in sieve elements whether such elements contained virus particles or not, and in parenchyma cells they occurred only in association with virus particles. The origin of the vesicles has not been determined as yet. They may be assumed to originate from some normal component of the infected cell but transitional stages in vesicle development that might relate them to a specific cell component were not recognized. Among the abnormal structures reported for virus-infected cells in the literature, the aggregates of hypertrophied dictyosomes and associated vesicles described by Rubio-Huertos (1965) in *Petunia* ringspot virus infection bears a comparison with vesicle aggregates in BYV-infected leaves. Some similarity is also detectable between these vesicle accumulations and the proliferations of endoplasmic reticulum observed by Gerola, Bassi, and Belli (1966) in *Petunia* infected with *Arabis* mosaic virus. A positive identification of the vesicles requires a developmental study.

Figure 125 shows a second kind of degenerated component of obscure origin: flakes of electron-opaque material (*F*). The common association of this material with ribosomes (Figs. 94, 95, and 125) could be interpreted as indicating origin from ribosomes. Cells approaching necrosis, however, show large complements of granules resembling ribosomes. If ribosomes degenerate, they probably do so only partially and not necessarily in all infected cells.

The phloem-parenchyma cell in Figure 126 shows a variant in the appearance of a degenerating protoplast. The main mass of cytoplasm contains few virus particles and the considerably degenerated nucleus includes no virions in this section. Most of the cytoplasmic virions are aggregated in the body of material in the center of the cell. The virions are intermixed with vesicles of the kind illustrated in Figure 125 and with ribosomes combined with the flaky material. Conceivably, an aggregation of this sort may appear as an amorphous inclusion body when examined with the light microscope. Flaky material is conspicuous also to the left of the nucleus. The chloroplasts are partly degenerated but the mitochondria are intact.

In later stages of degeneration the protoplasts of phloem-parenchyma cells show further increase in density (Fig. 127). Most organelles and membrane systems become disorganized and are replaced by electron-opaque material. Densely packed granules resembling ribosomes occur in this material. Broken-down chloroplasts release the starch grains. The vesicles that formerly appeared in small groups now form homogeneous aggregates, and their contents largely disappear. Masses of virus particles and irregular lumps of osmiophilic material are intermixed with the degenerated cell contents. The virus aggregates continue to be discernible in cells in which the host-cell components are completely necrosed (cf. Cronshaw et al., 1966), but cells exemplified by that bearing the number *1* in Figure 123 suggest that eventually the virus undergoes necrosis in dead cells. One might draw a parallel between this course of events and the hypersensitive reaction of cells in localized lesions with which certain hosts respond to viral infections. The rapid death of infected cells in the lesions may lead to the destruction of the virus and thus prevent its further spread. (The so-called resistant zone of apparently high metabolic activity surrounding a lesion according to Israel and Ross, 1967, is comparable to a wound-healing tissue.) In infections of the type induced by the beet yellows virus, cell necrosis probably destroys only a portion of the viral supply and does not prevent the continued spread of the virus to the growing regions of the plant. The plant itself does not necessarily die from viral infection; but as its metabolic activity is diverted toward the manufacture of virus and as the cells concerned with such vital functions as photosynthesis and translocation of food suffer degenerative changes, the normal functions of the plant decline and its growth is depressed.

The foregoing description of abnormalities in the phloem of sugar-beet leaves infected with beet yellows virus leads to the conclusion that the chromophilic cells visible with the light microscope in the neighborhood of mature sieve elements are indeed cells that have become invaded by the virus. This evidence corroborates the view presented earlier that the food conduit serves in the dispersal of a virus during the development of systemic infection and that, as exemplified by the beet yellows virus, the entities being dispersed may be complete virus particles.

The X-material and the P-protein

As was reviewed at the beginning of this chapter, analyses of viral effects on the host cells have led to the concept that these effects are of two distinct categories. First, virus directs the synthetic activity of the cell toward the production of virus. Second, virus brings into the host additional genetic information which may adversely affect the metabolism of the cell and cause degenerative changes. If this categorizing of viral effects is accepted, the abnormalities described in the preceding pages could serve as examples of the second kind of effects, changes resulting from disturbed metabolism. Figure 128, on the other hand, may be used to illustrate the result of the first category of viral influence. In this young mesophyll cell of tobacco the tobacco mosaic virus had induced the production of a large amount of virus and virus-related X-material without causing any discernible degenerative changes in the host protoplast.

The synthesis of viral components is a problem in biochemical genetics and its discussion lies outside the scope of this presentation. One of the aspects of viral multiplication, the production of virus-related protein, however, is of considerable cytologic interest to the student of the phloem tissue because of the similarity between the viral product and the protein (P-protein) normally synthesized by sieve elements and some of its associated cells. The sequence of development of the X-material in TMV infections—beginning with a granular viroplasmlike component, through complexes of tubules in circumscribed X-bodies, to individual tubules in aggregates within the cytoplasm—has a certain parallel in the development of the P-protein. As determined for the P-protein in *Cucurbita* (Cronshaw and Esau, 1968*b*) this material also becomes first visible as a precursory component, possibly as small fibrils, that is later transformed into tubules of more than one form. The tubules are initially in compact aggregates (slime bodies) but later become dispersed. For orientation, Figure 128 may be used for recalling the appearance of the X-material in the form of an X-body in juxtaposition with aggregates of complete virus particles, whereas Figure 129 exemplifies the P-protein in a young sieve element of tobacco, in which the nucleus is still intact and the vacuoles are bound by tonoplasts. The similarity between the X-tubules and the P-protein tubules is evident in the unusual view in Figure 130 in which the X-material and the P-protein happened to be present in the same section of one phloem-parenchyma cell.

When seen separately in different cells, as is mostly the case, the X-material and the P-protein are difficult to distinguish. The features differentiating the two components as tentatively pointed out by Esau and Cronshaw (1967b) were size, the X-tubules being somewhat wider than the P-protein tubules (approximately 280 A vs. 230 A), and the more orderly arrangement and straighter form of the X-tubules. The authors also considered it to be helpful that X-material was not seen in sieve elements, the main P-protein-producing cells. Subsequent studies, however, indicated that the arrangement and form of tubules were not consistently different—both kinds of tubule are straight and their arrangement is orderly in Figure 130—and X-material in all forms was found in the sieve elements (see Chapter 3). The size difference, though small, seems to be consistent and is obvious without a measurement when the two components occur side by side in the same cell (Fig. 130). A feature not previously described for the P-protein in tobacco is the occurrence of rows of granules between the tubules when these are still in more or less compact aggregates (Fig. 131). The granules are smaller than the ribosomes in the cytoplasm. They are still present when the P-protein bodies begin to disperse (Fig. 132). Similar granules were not seen among the X-tubules.

If the presence of granules will prove to be a reliable characteristic distinguishing between the X-material and the P-protein, it may assist in elucidating the significance of the peculiarly large accumulations of tubular material encountered in some phloem-parenchyma cells of tobacco leaves infected with TMV. The occurrence of P-protein in parenchyma cells more or less closely associated with the sieve elements is now a well-documented phenomenon. Normally the P-protein is encountered in phloem-parenchyma cells rather infrequently and in small amounts (Figs. 133 and 134). In some young infected leaves, however, the tubular aggregates in parenchyma cells were remarkably voluminous (Fig. 135) and occurred in more cells than in healthy plants. Though the tubules were arranged generally parallel to one another, the groups of tubules were twisted in various planes so that sections of the aggregates had highly complex patterns (Fig. 136). Presence of granules in rows among the tubules suggested that the material was P-protein rather than X-material. Moreover, the width of the tubules (Fig. 137) was similar to that observed in P-protein tubules in other views (Fig. 72). If the interpretation of the massive accumulations of tubules in the parenchyma cells as those of P-protein is correct, and the phenomenon will prove to be common in TMV infections, the matter of overproduction of P-protein should be examined with reference to the virus-directed synthesis. Tentatively, one can do no more than raise the question as to whether the virus-induced production of surplus viral protein has any relation to the synthesis of unusually large amounts of the native P-protein.

RETROSPECT

VIRUSES are frequently found assuming crystalline or paracrystalline forms in the cells of their plant hosts. This tendency has been noted in infections with both the isometric and the anisometric viruses. Whether the virus must reach a certain concentration in the cell before the particles become symmetrically aligned is not certain. Perhaps it suffices when the appropriate concentration is attained in a given region of the cell. Viral nucleoproteins share their tendency toward ordered arrangement with various proteins of the host cell and, judged by the situation in infections with tobacco mosaic virus, also with the RNA-free virus-related protein. Whether they occur in crystalline arrays or are randomly scattered, the virions may be so numerous in a cell that they appear to saturate the cytoplasm. At the same time, the distribution of the virus in a tissue is highly erratic. Cells filled with virus may occur side by side with those that are free of any microscopically discernible components of the virus. The inconsistency in viral distribution extends to the conducting cells. Some sieve elements are filled with virus particles, others contain little virus or none at all. In phloem-related viruses the presence or absence of discernible virus in the sieve elements apparently has no relation to the degree of degeneration of the cells associated with the sieve elements. The delivery of one or a few virions is probably sufficient to start an infection and the resulting degeneration in the parenchyma cells.

The cytoplasm is the most common site of virus particles in plant cells. Their occasional occurrence in vacuoles is suspected of being an artifact. Among the organelles, the nucleus is mentioned most often as containing virus. Particles of tobacco mosaic virus (but not those of any other virus so far) have been found in plastids, and not only in chloroplasts but in every kind of plastid present in the plant, including the very little differentiated sieve-element plastid. Particles of some viruses occur in the endoplasmic reticulum, but none was reported in dictyosomes, and they appear to be rather consistently absent from mitochondria. The occurrence of complete virus particles in organelles has been repeatedly discussed with reference to the problem of sites of viral synthesis, but the significance of the relation of the virus to the organelles continues to be problematic.

A comparative view of the changes induced by viruses in the host cells demonstrates that the symptoms vary in kind and in degree of severity. Some are of the kind that could be induced by a variety of agents, others are more specific. The severity of symptoms is not necessarily correlated with the amount of virus in the cell as determined by the presence of virions in the protoplast. Infections with tobacco mosaic virus furnish illustrations that large amounts of virus in the cell may have no apparent effect on the condition of the host-cell components, even to the extent that mitosis and cytokinesis are continued without obvious abnormalities. Viruses, of course, differ in their effects on the host. In infections with the beet yellows virus, virus-containing cells undergo necrosis. Among the organelles, the chloroplasts are most often reported as being adversely affected by viral infections.

With reference to the concept that virus brings only genetic information to the host cell and that all viral material is the product of the host cell, the similarity between the viral protein (X-material) in tobacco mosaic virus infection and the P-protein normally produced by certain phloem cells is of considerable interest. The components of both proteins are tubules (at least they have electron-lucent cores) and although the two kinds of tubules differ somewhat in width, they have common characteristics that differentiate them from the plant-cell microtubules. Plant cells produce a variety of tubular proteins; the viral protein is one of the variants. The formation of X-material does not appear to interfere with the production of P-protein in the same cell. This circumstance may have a bearing on the concept that the induction of viral multiplication must be differentiated from those viral effects which cause partial or complete degeneration of the host cell.

BIBLIOGRAPHY

Arnott, H. J., and K. M. Smith. 1967. Electron microscopy of virus-infected sunflower leaves. J. Ultrastructure Res. 19: 173–195.
———. 1968. Electron microscopic observations on the apparent replication in vivo of a plant virus. Virology 34: 25–35.
Atkinson, P. H., and R. E. F. Matthews. 1967. Distribution of tobacco mosaic virus in systemically infected tobacco leaves. Virology 32: 171–173.
Bailey, J. L., and A. G. Whyborn. 1963. The osmiophilic globules in chloroplasts. II: Globules of spinach-beet chloroplast. Biochem. Biophys. Acta 78: 163–174.
Bancroft, J. B., G. J. Hills, and R. Markham. 1967. A study of the self-assembly process in a small spherical virus. Formation of organized structures from protein subunits in vitro. Virology 31: 354–379.
Bawden, F. C. 1964. Plant viruses and virus diseases. 4th ed. Ronald Press Company, New York.
Bennett, C. W. 1940. The relation of viruses to plant tissues. Bot. Rev. 6: 427–473.
———. 1956. Biological relations of plant viruses. Annu. Rev. Plant Physiol. 7: 143–170.
Brandes, J. 1955. Elektronenmikroskopischer Nachweis von Pflanzenviren in ihren Wirtszellen. Naturwissenschaften 42: 101.
———. 1956. Über das Aussehen und die Verteilung des Tabakmosaikvirus im Blattgewebe; elektronenmikroskopische Un-

tersuchungen an ultradünnen Mikrotomschnitten. Phytopathol. Z. 26: 93–106.
Brandes, J., and C. Wetter. 1959. Classification of elongated plant viruses on the basis of particle morphology. Virology 8: 99–115.
Butler, R. D. 1967. The fine structure of senescing cotyledons of cucumber. J. Exp. Bot. 18: 535–543.
Carroll, T. W. 1966. Lesion development and distribution of tobacco mosaic virus in *Datura stramonium*. Phytopathology 56: 1348–1353.
Catesson, A.-M. 1966. Présence de phytoferritine dans le cambium et les tissus conducteurs de la tige de Sycomore, *Acer pseudoplatanus*. Compt. Rend. Acad. Sci. 262: 1070–1073.
Chalcroft, J., and R. E. F. Matthews. 1966. Cytological changes induced by turnip yellow mosaic virus in Chinese cabbage leaves. Virology 28: 555–562.
Chambers, T. C., and R. I. B. Francki. 1966. Localization and recovery of lettuce necrotic yellows virus from xylem tissues of *Nicotiana glutinosa*. Virology 29: 673–676.
Chandra, N., and A. C. Hildebrandt. 1966. Growth in microculture of single tobacco cells infected with tobacco mosaic virus. Science 152: 789–791.
———. 1967. Differentiation of plants from tobacco mosaic virus inclusion-bearing and inclusion-free single tobacco cells. Virology 31: 414–421.
Commoner, B. 1959. The biochemistry of the synthesis and biological activity of tobacco mosaic virus. *In* C. S. Holton et al., Plant Pathology: Problems and Progress, 1908–1958, pp. 483–492. The University of Wisconsin Press, Madison, Wisconsin.
Cronshaw, J., and K. Esau. 1968a. Cell division in leaves of *Nicotiana*. Protoplasma 65: 1–24.
———. 1968b. P-protein in the phloem of *Cucurbita*. I: The development of P-protein bodies. J. Cell Biol. 38: 25–39.
Cronshaw, J., L. L. Hoefert, and K. Esau. 1966. Ultrastructural features of *Beta* leaves infected with beet yellows virus. J. Cell Biol. 31: 429–443.

De Zoeten, G. A. 1966. California tobacco rattle virus, its intracellular appearance, and the cytology of the infected cell. Phytopathology 56: 744–754.

De Zoeten, G. A., and D. E. Schlegel. 1967a. Broadbean mottle virus in leaf tissue. Virology 31: 173–176.

———. 1967b. Nucleolar and cytoplasmic uridine-H^3 incorporation in virus infected plants. Virology 32: 416–427.

Diener, T. O. 1963. Physiology of virus-infected plants. Annu. Rev. Phytopathol. 1: 197–218.

Edwardson, J. R. 1966. Electron microscopy of cytoplasmic inclusions in cells infected with rod-shaped viruses. Amer. J. Bot. 53: 359–364.

Edwardson, J. R., D. E. Purcifull, and R. G. Christie. 1968. Structure of cytoplasmic inclusions in plants infected with rod-shaped viruses. Virology 34: 250–263.

Engelbrecht, A. H. P., and K. Esau. 1963. Occurrence of inclusions of beet yellows viruses in chloroplasts. Virology 21: 43–47.

Esau, K. 1941. Inclusions in guard cells of tobacco affected with mosaic. Hilgardia 13: 427–434.

———. 1948. Some anatomical aspects of plant virus disease problems, II. Bot. Rev. 14: 413–449.

———. 1958. Phloem degeneration in celery infected with yellow leafroll virus of peach. Virology 6: 348–356.

———. 1960a. Cytologic and histologic symptoms of beet yellows. Virology 10: 73–85.

———. 1960b. The development of inclusions in sugar beets infected with the beet-yellows virus. Virology 11: 317–328.

———. 1961. Plants, viruses, and insects. Harvard University Press, Cambridge, Massachusetts.

———. 1965. Fixation images of sieve element plastids in *Beta*. Nat. Acad. Sci. Proc. 54: 429–437.

———. 1966. Explorations of the food conducting system in plants. Amer. Sci. 54: 141–157.

———. 1967a. Minor veins in *Beta* leaves: structure related to function. Proc. Amer. Phil. Soc. 111: 219–233.

Esau, K. 1967b. Antomy of plant virus infections. Annu. Rev. Phytopathol. 5: 45–76.

Esau, K., and J. Cronshaw. 1967a. Relation of tobacco mosaic virus to the host cells. J. Cell Biol. 33: 665–678.

———. 1967b. Tubular components in cells of healthy and tobacco mosaic virus-infected *Nicotiana*. Virology 33: 26–35.

Esau, K., J. Cronshaw, and L. L. Hoefert. 1966. Organization of beet yellows-virus inclusions in leaf cells of *Beta*. Nat. Acad. Sci. Proc. 55: 486–493.

———. 1967. Relation of beet yellows virus to the phloem and to translocation in the sieve tube. J. Cell Biol. 32: 71–87.

Evert, R. F., and W. F. Derr. 1964. Slime substance and strands in sieve elements. Amer. J. Bot. 51: 875–880.

Feldherr, C. M. 1962. The nuclear annuli as pathways for nucleocytoplasmic exchanges. J. Cell Biol. 14: 65–72.

———. 1965. The effect of the electron-opaque pore material on exchanges through the nuclear annuli. J. Cell Biol. 25: 43–53.

Fujisawa, I., T. Hayashi, and C. Matsui. 1967. Electron microscopy of mixed infections with two plant viruses. I: Intracellular interactions between tobacco mosaic virus and tobacco etch virus. Virology 33: 70–76.

Fukushi, T., and E. Shikata. 1963. Localization of rice dwarf virus in its insect vector. Virology 21: 503–505.

Fukushi, T., E. Shikata, and I. Kimura. 1962. Some morphological characters of rice dwarf virus. Virology 18: 192–205.

Gardner, W. S. 1967. Electron microscopy of barley stripe mosaic virus: comparative cytology of tissues infected during different stages of maturity. Phytopathology 57: 1315–1326.

Gerola, F. M., and M. Bassi. 1966. An electron microscopy study of leaf vein tumors from maize plants experimentally infected with maize rough dwarf virus. Caryologia 19: 13–40.

Gerola, F. M., M. Bassi, and G. G. Belli. 1966. Some observations on the shape and localization of different viruses in experimentally infected plants, and on the fine structure of the host cells. IV: Arabis mosaic virus in *Petunia hybrida* Hort. Caryologia 19: 481–491.

Gerola, F. M., M. Bassi, and G. Giussani. 1966. Some observations on the shape and localization of different viruses in experimentally infected plants, and on the fine structure of the host cells. III: Turnip yellow mosaic virus in *Brassica chinensis* L. Caryologia 19: 457–479.

Gerola, F. M., M. Bassi, O. Lovisolo, and C. Vidano. 1966. Electron microscopic observations on wheat plants experimentally infected with maize rough dwarf virus. Caryologia 19: 493–503.

Gol'din, M. I., N. V. Agoeva, and V. A. Tumanova. 1967. Prizhiznennoe issledovanie virusnykh vklyuchenii kazakhskogo shtamma virusa mozaiki tabaka v kul'ture tkani i otdel'nykh kletok. Acta Virol. 11: 462–463.

Goldstein, B. 1926. A cytological study of the leaves and growing points of healthy and mosaic diseased tobacco plants. Bull. Torrey Bot. Club 53: 499–599.

Greenwood, A. D., R. M. Leech, and J. P. Williams. 1963. The osmiophilic globules of chloroplasts. I: Osmiophilic globules as normal component of chloroplasts and their isolation and composition in *Vicia faba* L. Biochem. Biophys. Acta 78: 148–162.

Hepler, P. K., and E. H. Newcomb. 1967. Fine structure of cell plate formation in the apical meristem of *Phaseolus* roots. J. Ultrastructure Res. 19: 498–513.

Hiebert, E., J. B. Bancroft, and C. E. Bracker. 1968. The assembly in vitro of some small spherical viruses, hybrid viruses, and other nucleoproteins. Virology 34: 492–508.

Hirai, A., and S. G. Wildman. 1967. Intracellular site of assembly of TMV-RNA and protein. Virology 33: 467–473.

Hitchborn, J. H., and G. J. Hills. 1967. Tubular structures associated with turnip yellow mosaic virus in vivo. Science 157: 705–706.

Hoefert, L. L., and K. Esau. 1967. Degeneration of sieve element plastids in sugar beet infected with curly top virus. Virology 31: 422–426.

Holter, H. 1961. Pinocytosis. *In* T. W. Goodwin and O. Lindberg, eds., Biological Structure and Function, pp. 157–168. Academic Press, New York.

Hyde, B. B., A. J. Hodge, and M. L. Birnstiel. 1962. Phytoferritin: a plant protein discovered by electron microscopy. *In* Fifth International Congress for Electron Microscopy, Vol. 2, p. T-1. Academic Press, New York.

Hyde, B. B., A. J. Hodge, A. Kahn, and M. L. Birnstiel. 1963. Studies on phytoferritin. I: Identification and localization. J. Ultrastructure Res. 9: 248–258.

Israel, H. W., and A. F. Ross. 1967. The fine structure of local lesions induced by tobacco mosaic virus in tobacco. Virology 33: 272–286.

Karnovsky, M. J. 1965. A formaldehyde-glutaraldehyde fixative of high osmolality for use in electron microscopy. J. Cell Biol. 27: 137A–138A.

Kausche, G. A., E. Pfankuch, and H. Ruska. 1939. Die Sichtbarmachung von pflanzlichem Virus mit Übermikroskop. Naturwissenschaften 27: 292.

Kitajima, E. W. 1965. Electron microscopy of vira-cabeça virus (Brazilian tomato spotted wilt virus) within the host cell. Virology 26: 89–99.

Kitajima, E. W., D. M. Silva, A. R. Oliveira, G. W. Müller, and A. S. Costa. 1964. Thread-like particles associated with tristeza disease of citrus. Nature 201: 1011–1012.

Knight, C. A. 1963. Chemistry of viruses. Protoplasmatologia 4: 1–177.

Kolehmainen, L., H. Zech, and D. von Wettstein. 1965. The structure of cells during tobacco mosaic virus reproduction. Mesophyll cells containing virus crystals. J. Cell Biol. 25 (3, pt. 2) : 77–97.

Lee, P. E. 1967. Morphology of wheat striate mosaic virus and its localization in infected cells. Virology 33: 84–94.

Lichtenthaler, H. K. 1966. Plastoglobuli und Plastidenstruktur. Ber. Deut. Bot. Ges. 79: (82)–(88).

Luria, S. E. 1958. The multiplication of viruses. Protoplasmatologia 4: 1–63.

López-Sáez, J. F., G. Giménez-Martín, and M. Risueño. 1966. Fine structure of the plasmodesm. Protoplasma 61: 81–84.

Milne, R. G. 1966. Multiplication of tobacco mosaic virus in tobacco leaf palisade cells. Virology 28: 79–89.
———. 1967a. Electron microscopy of leaves infected with sowbane mosaic virus and other small polyhedral viruses. Virology 32: 589–600.
———. 1967b. Plant viruses inside cells. Sci. Progress, Oxford, 55: 203–222.
Mundry, K. W. 1963. Plant virus-host cell relations. Annu. Rev. Phytopathol. 1: 173–196.
Nakata, K., and A. C. Hildebrandt. 1967. Fine structure of healthy and TMV-infected tobacco cells grown in tissue culture. Physiol. Pl. 20: 999–1013.
Newcomb, E. H. 1967. Fine structure of protein-storing plastids in bean root tips. J. Cell Biol. 33: 143–163.
O'Brien, T. P. 1967. Observations on the fine structure of the oat coleoptile. I: The epidermal cells of the extreme apex. Protoplasma 63: 385–416.
O'Brien, T. P., and K. V. Thimann. 1967. Observations on the fine structure of the oat coleoptile. II: The parenchyma cells of the apex. Protoplasma 63: 417–442.
Peveling, E., and H. K. Lichtenthaler. 1967. Zur Funktion der osmiophilen Lipideinschlüsse im Cytoplasma der Zwiebelschalen von *Allium cepa* L. Z. Pflanzenphysiol. 56: 299–303.
Price, W. C. 1966. Flexuous rods in phloem cells of lime plants infected with citrus tristeza virus. Virology 29: 285–294.
Purcifull, D. E., and J. R. Edwardson. 1967. Watermelon mosaic virus: tubular inclusions in pumpkin leaves and aggregates in leaf extracts. Virology 32: 393–401.
Ragetli, H. W. J. 1967. Virus-host interactions, with emphasis on certain cytoplasmic phenomena. Can. J. Bot. 45: 1221–1234.
Reddi, K. K. 1966. Studies on the formation of tobacco mosaic virus ribonucleic acid. VII: Fate of tobacco mosaic virus after entering the host cell. Nat. Acad. Sci. Proc. 55: 593–598.
Robards, A. W., and P. G. Humpherson. 1967. Phytoferritin in plastids of the cambial zone of willow. Planta 76: 169–178.
Rubio-Huertos, M. 1965. Golgi apparatus hypertrophy associated

with *Petunia* ringspot virus infection. Microbiol. Espan. 18: 195–202.

Rubio-Huertos, M., and A. Vela-Cornejo. 1966. Light and electron microscopy of virus inclusions in *Amaranthus lividus* cells. Protoplasma 62: 184–193.

Russell, G. E., and J. Bell. 1963. The structure of beet yellows virus filaments. Virology 21: 283–284.

Russo, M., G. P. Martelli, and A. Quacquarelli. 1967. Occurrence of artichoke mottled crinkle virus in leaf vein xylem. Virology 33: 555–558.

———. 1968. Studies on the agent of artichoke mottled crinkle. IV: Intracellular localization of the virus. Virology 34: 679–693.

Sanders, F. K. 1964. Viruses and cells. *In* G. H. Bourne, ed., Cytology and Cell Physiology, Chap. 11, pp. 637–665. Academic Press, New York.

Schlegel, D. E., S. H. Smith, and G. A. de Zoeten. 1967. Sites of virus synthesis within cells. Annu. Rev. Phytopathol. 5: 223–246.

Schneider, I. R. 1965. Introduction, translocation, and distribution of viruses in plants. Advance. Virus Res. 11: 163–221.

Schneider, I. R., and J. F. Worley. 1959a. Upward and downward transport of infectious particles of southern bean mosaic through steamed portions of bean stems. Virology 8: 230–242.

———. 1959b. Rapid entry of infectious particles of southern bean mosaic virus into living cells following transport of the particles in the water stream. Virology 8: 243–249.

Schnepf, E. 1961. Plastidenstrukturen bei *Passiflora*. Protoplasma 54: 310–313.

Schramm, G. 1961. Biosynthesis of virus proteins. *In* R. J. C. Harris, ed., Protein Biosynthesis, pp. 337–344. Academic Press, New York.

Shalla, T. A. 1964. Assembly and aggregation of tobacco mosaic virus in tomato leaflets. J. Cell Biol. 21: 253–264.

Shaw, J. G. 1967. In vivo removal of protein from tobacco mosaic virus after inoculation of tobacco leaves. Virology 31: 665–675.

Sheffield, F. M. L. 1936. The roll of plasmodesms in the translocation of virus. Ann. Appl. Biol. 23: 506–508.

Shikata, E., and K. Maramorosch. 1965. Electron microscopic evidence for the systemic invasion of an insect host by a plant pathogenic virus. Virology 27: 461–475.

———. 1966a. An electron microscopy of plant neoplasia induced by wound tumor virus. J. Nat. Cancer Inst. 36: 97–116.

———. 1966b. Electron microscopy of pea enation mosaic virus in plant cell nuclei. Virology 30: 439–454.

———. 1967. Electron microscopy of wound tumor virus assembly sites in insect vectors and plants. Virology 32: 363–377.

Shikata, E., K. Maramorosch, and R. R. Granados. 1966. Electron microscopy of pea enation mosaic virus in plants and aphid vectors. Virology 29: 426–436.

Silaeva, A. M., and A. T. Shiryaev. 1966. O tonkoi strukture osmiofil'nykh globul khloroplasta. Dokl. Akad. Nauk SSSR 170: 433–434.

Singh, M., and A. C. Hildebrandt. 1966. Movements of tobacco mosaic virus inclusion bodies within tobacco callus cells. Virology 30: 134–142.

Sitte, P. 1961. Zum Bau der Plastidenzentren in Wurzelproplastiden. Protoplasma 53: 438–442.

———. 1965. Bau und Feinbau der Pflanzenzelle. Gustav Fischer Verlag, Stuttgart.

Sprey, B. 1965. Beiträge zur makromolekularen Organization der Plastiden I. Z. Pflanzenphysiol. 53: 255–261.

Steere, R. L., and R. C. Williams. 1953. Identification of crystalline inclusion bodies extracted from plant cells infected with tobacco mosaic virus. Amer. J. Bot. 40: 81–84.

Strugger, S. 1957. Schraubig gewundene Fäden als sublichtmikroskopische Strukturelemente des Cytoplasmas. Ber. Deut. Bot. Ges. 70: 91–108.

Strunk, C., and H. Wartenberg. 1960. Licht- und elektronenop-

tische Untersuchungen der Chloroplasten chlorotischer Maispflanzen. Phytopathol. Z. 38: 109–122.
Takahashi, W. N., and M. Ishii. 1953. A macromolecular protein associated with tobacco mosaic virus infection: its isolation and properties. Amer. J. Bot. 40: 85–90.
Thomson, W. W. 1966. Ultrastructural development of chromoplasts in Valencia oranges. Bot. Gaz. 127: 133–139.
Thomson, W. W., L. N. Lewis, and C. W. Coggins. 1967. The reversion of chromoplasts to chloroplasts in Valencia oranges. Cytologia 32: 117–124.
Thomson, W. W., T. E. Weier, and H. Drever. 1964. Electron microscopic studies on chloroplasts from phosphorus-deficient plants. Amer. J. Bot. 51: 933–938.
Warmke, H. E., and J. R. Edwardson. 1966a. Use of potassium permanganate as a fixative for virus particles in plant tissues. Virology 28: 693–700.
———. 1966b. Electron microscopy of crystalline inclusions of tobacco mosaic virus in leaf tissue. Virology 30: 45–57.
Wehrmeyer, W. 1959. Entwicklungsgeschichte, Morphologie und Struktur von Tabakmosaikvirus-Einschlusskörpern unter besonderer Berücksichtigung der fibrillären Formen. Protoplasma 51: 165–196.
Weintraub, M., and H. W. J. Ragetli. 1964a. Studies on the metabolism of leaves with localized virus infections. Particulate fractions and substrates in TMV-infected *Nicotiana glutinosa* L. Can. J. Bot. 42: 533–540.
———. 1964b. An electron microscopic study of tobacco mosaic virus lesions in *Nicotiana glutinosa* L. J. Cell Biol. 23: 499–509.
Weintraub, M., H. W. J. Ragetli. and M. M. Dwurazna. 1964. Studies on the metabolism of leaves with localized virus infections. Mitochondrial activity in TMV-infected *Nicotiana glutinosa* L. Can. J. Bot. 42: 541–545.
Weintraub, M., H. W. J. Ragetli, and V. T. John. 1967. Some conditions affecting the intracellular arrangement and concentration of tobacco mosaic virus particles in local lesions. J. Cell Biol. 35: 183–192.

Wettstein, D. von, and H. Zech. 1962. The structure of nucleus and cytoplasm in hair cells during tobacco mosaic virus reproduction. Z. Naturforsch. 17b: 376–379.

Wilkins, M. H. F., A. R. Stokes, W. E. Seeds, and G. Oster. 1950. Tobacco mosaic virus crystals and three-dimensional microscopic vision. Nature 166: 127–129.

Woods, M. W., and R. V. Eck. 1948. Nuclear inclusions produced by a strain of tobacco mosaic virus. Phytopathology 38: 852–856.

Zaitlin, M., and N. K. Boardman. 1958. The association of tobacco mosaic virus with plastids. I: Isolation of virus from the chloroplast fraction of diseased-leaf homogenates. Virology 6: 743–757.

Zech, H. 1960. Intermediary products formed during tobacco mosaic virus reproduction. Virology 11: 499–502.

ILLUSTRATIONS

ABBREVIATIONS

CA, callose
CC, companion cell
CH, chloroplast
CM, chromosome
CP, cell plate
D, dictyosome
ER, endoplasmic reticulum
F, flaky material
FR, fibrous ring
GR, granum
IS, intercellular space
K, kinetochore
M, mitochondrion
MB, membrane
ML, myelin formation
MT, microtubule
N, nucleus
NE, nuclear envelope
NP, nuclear pore
O, osmiophilic material

PA, parenchyma cell
PD, plasmodesma
PL, plastid
PM, plasmalemma
PP, P-protein
PR, protrusion
PW, primary wall
RB, ribosome
S, sieve element
SA, stomatal aperture
SP, sieve plate
ST, starch
SW, secondary wall
T, tubule
TO, tonoplast
V, virus
VA, vacuole
VE, vesicle
W, cell wall
X, X-material or X-body

NOTE ON TECHNIQUES

Figs. 6 and 14, light-microscope photomicrographs; all others, electron-microscope micrographs.

Fixations: potassium permanganate, Figs. 1–3, 64, and 91; osmium tetroxide, Fig. 73; glutaraldehyde-osmium tetroxide, Figs. 23–28, 34, 38, 63, 67, 68, 74–80, 89, 90, 92–110, 114–127; glutaraldehyde-paraformaldehyde-osmium tetroxide, Figs. 4, 5, 7–13, 15–22, 29–33, 35–37, 39–62, 65, 66, 69–72, 81–88, 111–113, 128–137.

Figs. 1–3. Crystalline arrangement of tobacco mosaic virus particles in sections of parenchyma cells of *Nicotiana tabacum* fixed with potassium permanganate.

Fig. 1 (*above*). Portion of crystal. Rod-shaped particles arranged parallel to each other form several layers. Inclined position of particles in some layers produces a herringbone pattern. Separation of layers from each other is probably an artifact. × 26,000.

Fig. 2 (*middle*). Single layer (monolayer) of virus particles. × 31,000.

Fig. 3 (*below*). Section of crystal at right angles to long axes of virus particles. Transections of rods are in orderly hexagonal arrangement. As in Figure 1, layers of particles are separated from one another. Identity of fibrous material between layers is not known. × 80,000.

Figs. 4–6. Forms of aggregates of tobacco mosaic virus particles in sections of parenchyma cells of *Nicotiana tabacum*.

Fig. 4 (*above*). Portion of crystal. Rod-shaped particles arranged parallel to each other form several layers. In places, host cytoplasm is enclosed between layers. × 30,000.

Fig. 5 (*middle*). Portion of aggregate composed of long rods. × 24,000.

Fig. 6 (*below*). Aggregate appearing as striate material in light-microscope view. × 1,500.

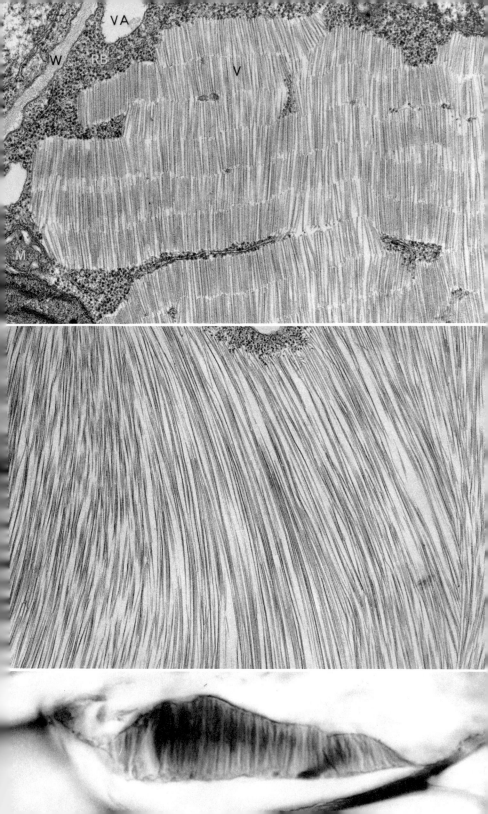

Figs. 7–9. Forms of aggregates of tobacco mosaic virus particles in sections of parenchyma cells of *Nicotiana tabacum.*

Fig. 7 (*above*). Prismatic aggregate of long rods embedded in cytoplasm. × 18,000.

Fig. 8 (*middle*). Massive aggregate, with long rods in two opposite orientations. Cytoplasm containing aggregate bulges out into vacuole but is delimited by tonoplast. × 21,000.

Fig. 9 (*below*). Needle-shaped aggregate of long rods embedded in cytoplasm. × 30,000.

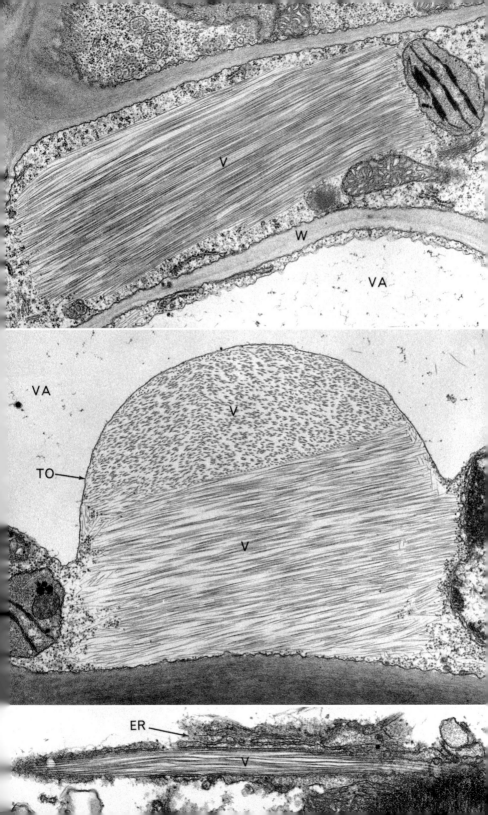

Figs. 10–12. Morphology of particles of tobacco mosaic virus in sections of parenchyma cells of *Nicotiana tabacum*.

Fig. 10 (*above*). Particles in transection reveal three regions: central electron-lucent canal; electron-opaque RNA-containing region around canal; outer protein-coat region of medium opacity. × 480,000.

Fig. 11 (*middle*). Particles in oblique and longitudinal sections. Protein coat less distinct than in transection. × 480,000

Fig. 12 (*below*). Aggregate with most of the particles of usual appearance: mainly electron-opaque core is discernible. In lower part of aggregate helically arranged component appears to cover particles. × 130,000.

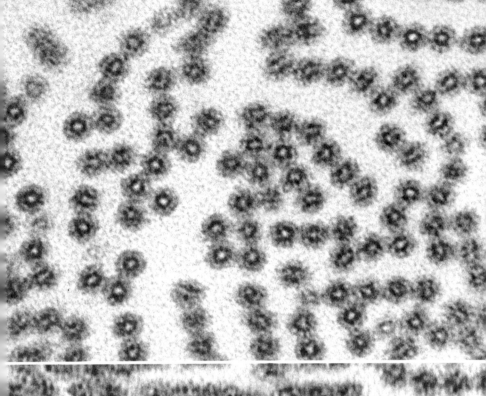
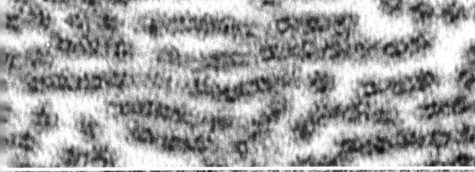
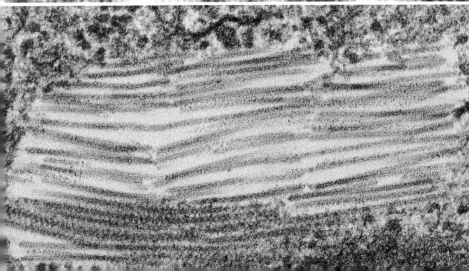

Fig. 13. Morphology of particles of tobacco mosaic virus in section of phloem-parenchyma cell of *Nicotiana tabacum*. At *V*, particles show usual form: mainly electron-opaque core is discernible. Elsewhere in the aggregate, particles appear to be covered with helically arranged component. × 67,500.

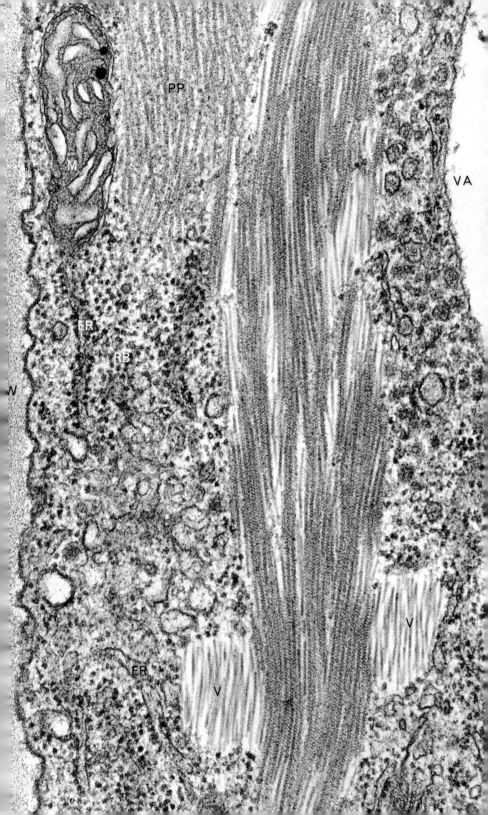

Figs. 14–16. X-material (virus-related protein) in sections of parenchyma cells of *Nicotiana tabacum* infected with tobacco mosaic virus.

Fig. 14 (*above, insert*). Light-microscope view of X-material aggregated into a vacuolate X-body. × 880.

Fig. 15 (*above, large picture*). X-body containing viral protein in form of bands disposed singly or in groups (*arrows*). Host-cell ribosomes occur among the bands. Some vacuoles and aggregates of virus particles are present. × 18,000.

Fig. 16. Portion of X-body with some protein bands in transection (*arrows*). Bands consist of tubules, usually in groups of three. × 82,500.

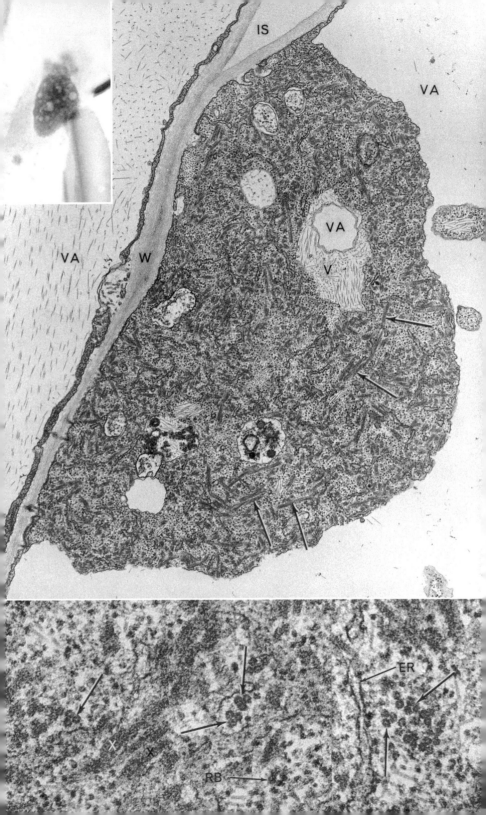

Figs. 17, 18. X-material in early stages of development in sections of parenchyma cells of *Nicotiana tabacum* infected with tobacco mosaic virus.

Fig. 17 (*above*). Mass of apparently granular material embedded in cytoplasm. × 45,000.

Fig. 18 (*below*). X-material, partly in granular form, partly as bands, embedded in cytoplasm. × 45,000.

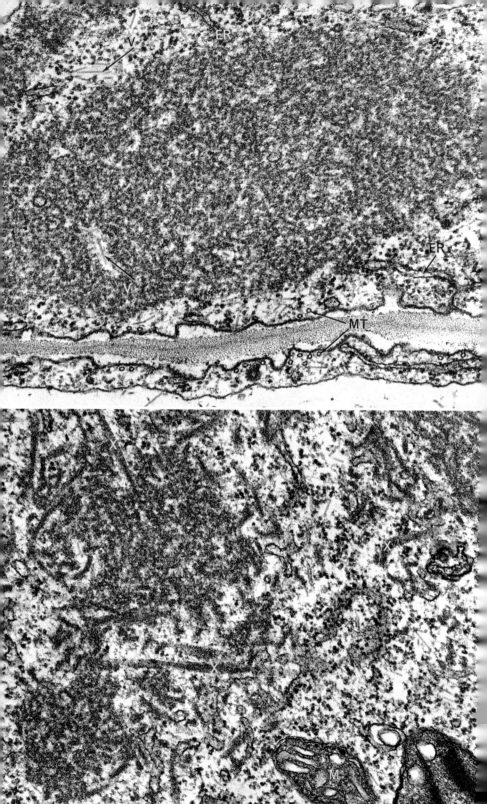

Figs. 19, 20. X-material in two advanced developmental stages in sections of parenchyma cells of *Nicotiana tabacum* infected with tobacco mosaic virus.

Fig. 19 (*above*). X-material in form of tubules which occur in groups of three (*arrows*) or singly. Most of the tubules are shown in transection. × 82,500.

Fig. 20 (*below*). X-material within cytoplasm in latest developmental stage. Tubules constitute individual components of small aggregates. Virus particles occur in the vacuole. X-tubules are considerably larger than the particles and contain no dark-staining core. (Compare with Fig. 10.) × 120,000.

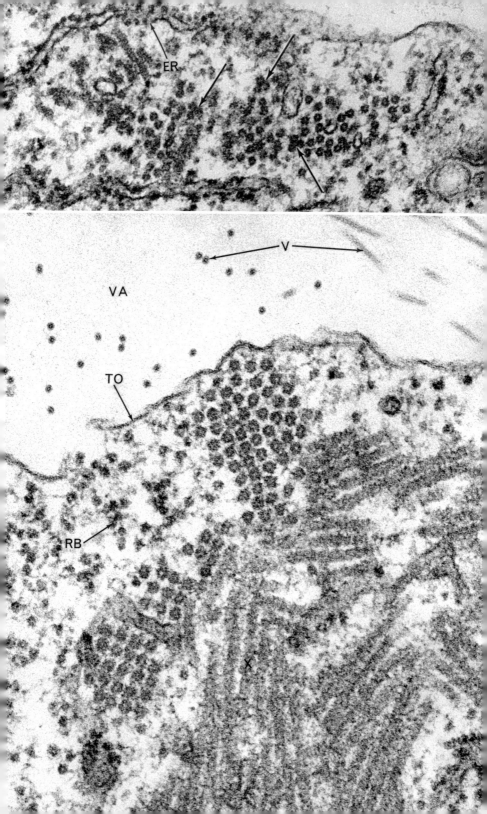

Fig. 21. Portion of an X-body in section of parenchyma cell of *Nicotiana tabacum* infected with tobacco mosaic virus. X-material at advanced stage of development. Tubules as individual components of small aggregates; in ordered arrangement at arrow. Ribosomes, cisternae of endoplasmic reticulum, groups of virus particles, and vacuoles occur in the X-body. × 45,000.

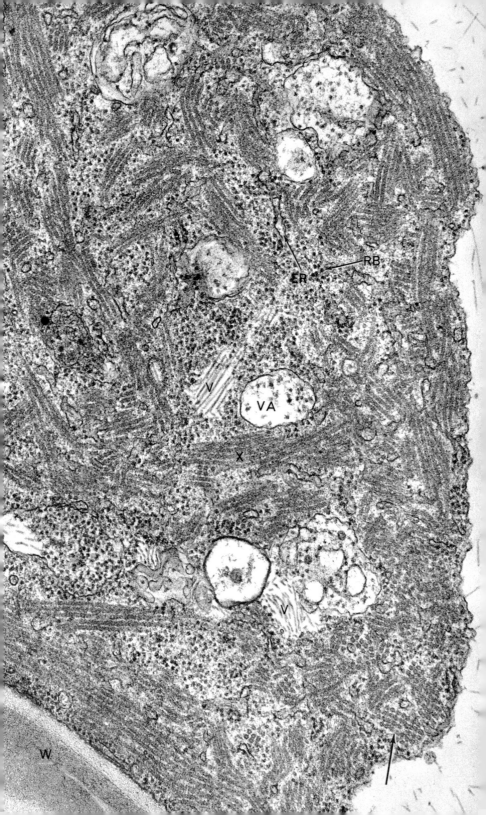

Fig. 22. X-material in form of bundle of long tubules in section of young parenchyma cell of *Nicotiana tabacum* infected with tobacco mosaic virus. Nucleus cut on a tangent shows pores in face view. Bias cut through wall has exposed microtubules. × 28,000.

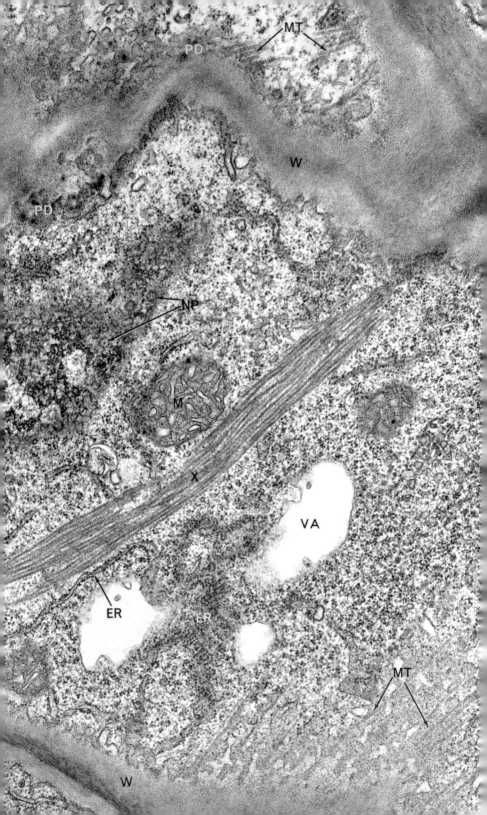

Figs. 23–25. Viral inclusions in sections of phloem-parenchyma cells in leaves of *Beta vulgaris* infected with beet yellows virus.

Fig. 23 (*above*). Banded inclusion composed of virus particles almost fills cell. × 3,750.

Fig. 24 (*middle*). Portion of banded inclusion. Bands are layers of long slender particles. × 30,000.

Fig. 25 (*below*). Portion of banded inclusion in transection. Virus particles in ordered arrangement in some areas. × 90,000.

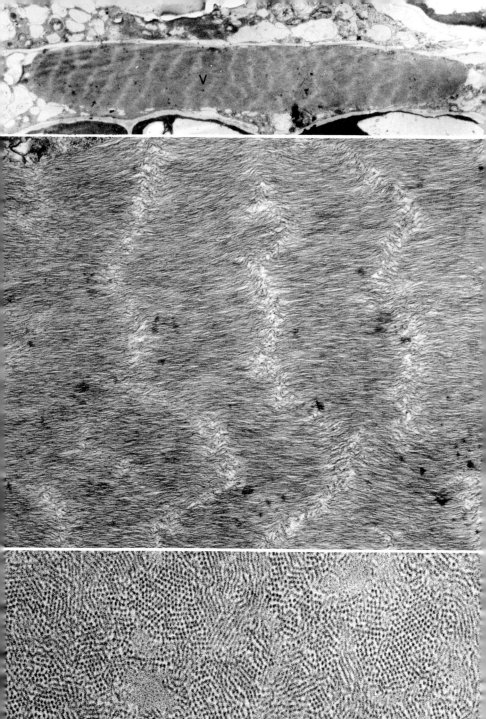

Figs. 26, 27. Viral inclusions in sections of phloem-parenchyma cells in leaves of *Beta vulgaris* infected with beet yellows virus.

Fig. 26 (*large picture*). Amorphous (nonbanded) inclusion composed of virus particles almost fills cell located next to sieve elements. × 8,000.

Fig. 27 (*below, insert*). Transection of nonbanded inclusion. Virus particles show slight tendency toward ordered arrangement. × 70,000.

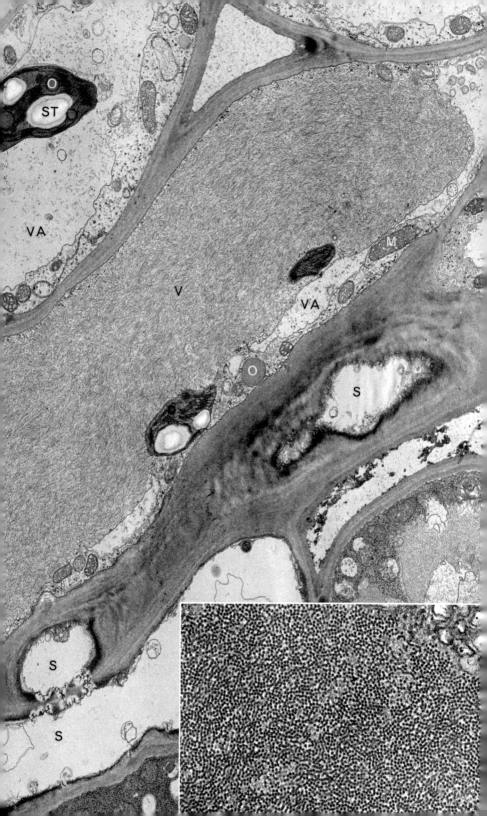

Fig. 28. Massive aggregation of particles of beet yellows virus in section of phloem-parenchyma cell from leaf of *Beta vulgaris*. Nucleus contains virus aggregate near nucleolus. In host cytoplasm, ribosomes, mitochondria, and plastids are interspersed with virus. Chloroplasts contain starch. Cell to the left, below, is free of virus. × 20,000.

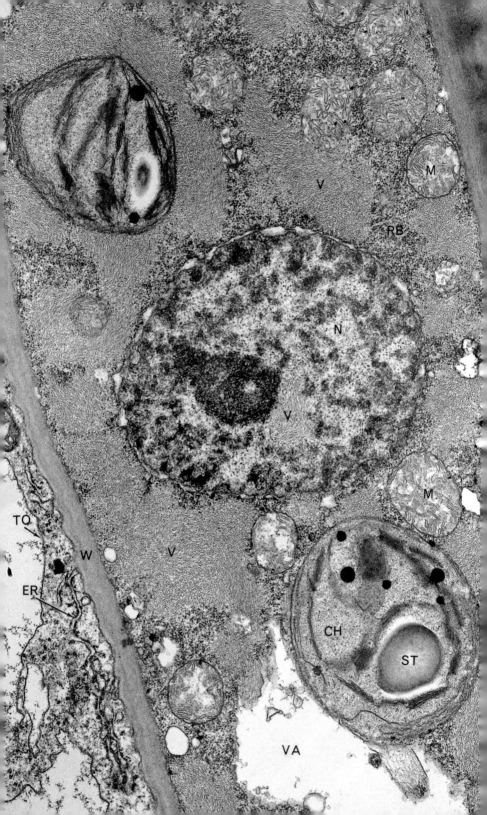

Fig. 29. Distribution of tobacco mosaic virus in section of parenchyma cell of *Nicotiana tabacum*. Virus occurs in cytoplasm and in one vacuole of cell in center. Break in tonoplast at arrow; possible passageway for virus from cytoplasm to vacuole. Protrusion from cytoplasm containing densely packed virus particles has encroached upon vacuole. At *PM*, invagination of plasmalemma into vacuole ("lomasome"). × 15,000.

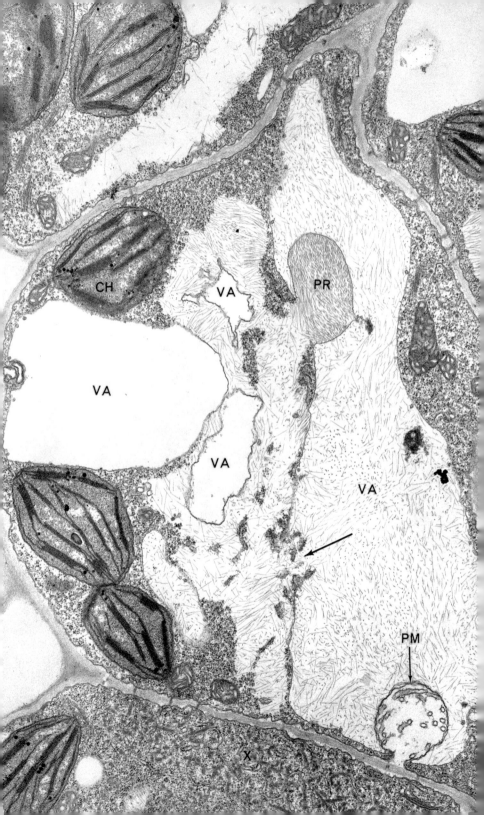

Figs. 30, 31. Distribution of tobacco mosaic virus in sections of parenchyma cells of *Nicotiana tabacum*.

Fig. 30 (*above*). Virus aggregate in parietal cytoplasm. Layering in aggregate indicates crystalline structure. Accumulation of virus has formed a bulge in cytoplasmic layer. Ribosome-rich cytoplasm occurs between wall and virus and between tonoplast and virus. × 18,000.

Fig. 31 (*below*). Virus aggregate in parietal cytoplasm occurs in pouchlike protrusion extending deeply into vacuole. Tonoplast separates virus from vacuole. × 30,000.

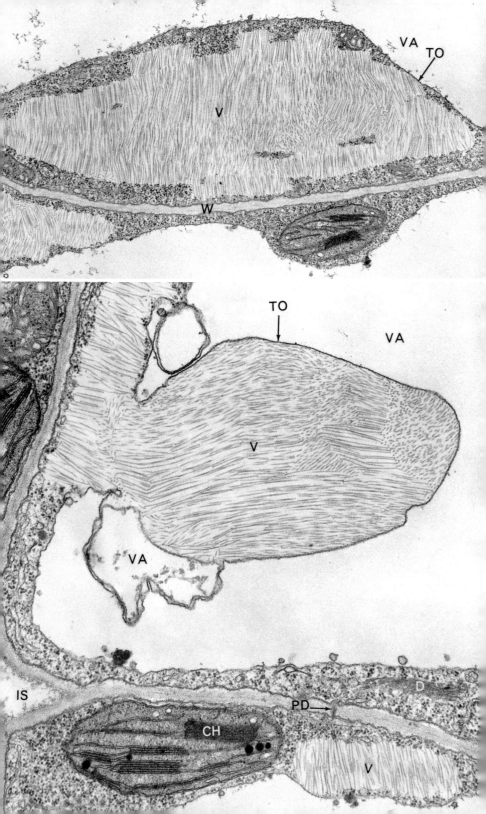

Figs. 32–35. Distribution of viruses in parenchyma tissue: tobacco mosaic virus in *Nicotiana tabacum* (Figs. 32, 33, and 35) and beet yellows virus in *Beta vulgaris* (Fig. 34).

Fig. 32 (*above*). Virus aggregate in bulge in cytoplasmic strand crossing vacuole. × 7,500.

Fig. 33 (*middle, left*). Protrusion filled with virus cut transversely. Tonoplast separates virus from surrounding vacuole. × 20,000.

Fig. 34 (*middle, right*). Protrusion with long beet yellows virus particles concentrically arranged around core of cytoplasm. Tonoplast separates virus from surrounding vacuole. × 37,500.

Fig. 35 (*below*). Crystalline aggregate of virus particles in intercellular space. × 25,000.

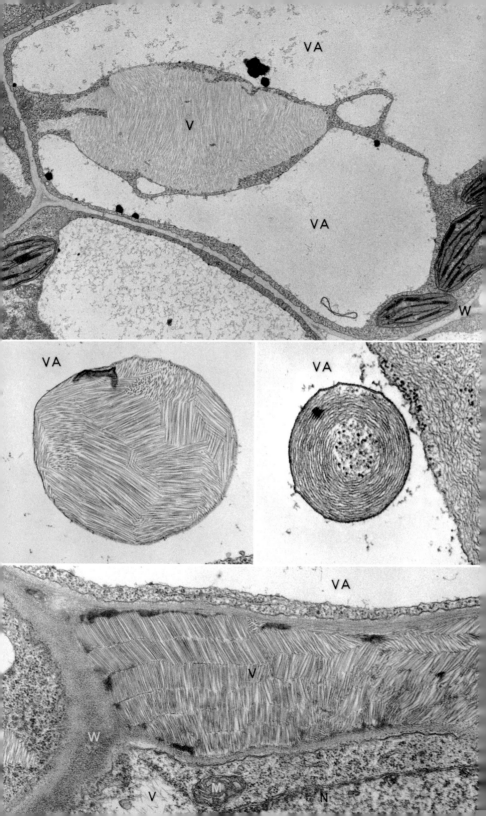

Figs. 36–38. Tobacco mosaic virus in relation to nuclei in parenchyma cells of *Nicotiana tabacum*.

Fig. 36 (*above*). Portion of nucleus with virus aggregate in center. × 16,500.

Fig. 37 (*below, large picture*). Portion of much-lobed nucleus to the left. Virus aggregates occur in invaginations in nucleus containing cytoplasm. Nuclear envelope delimits cytoplasm in invaginations. × 30,000.

Fig. 38 (*below, insert*). Fragment of nuclear envelope showing nuclear pores in face view. Several pores have a central dot. From noninfected plant. × 52,500.

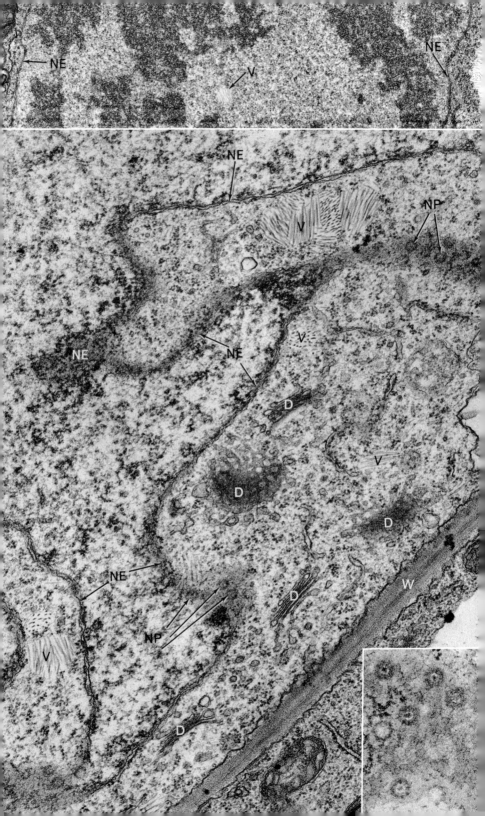

Figs. 39, 40. Tobacco mosaic virus in relation to dividing nuclei in parenchyma cells of *Nicotiana tabacum*.

Fig. 39 (*above*). Nucleus in metaphase. At two opposite poles, two arrays of endoplasmic-reticulum cisternae derived, at least in part, from nuclear envelope. Virus aggregate occurs among the cisternae. Metaphase chromosomes at *CM*, X-material at *X*. × 12,000.

Fig. 40 (*below*). Virus aggregate near metaphase chromosomes. Microtubules are attached to kinetochore to the right. × 24,000.

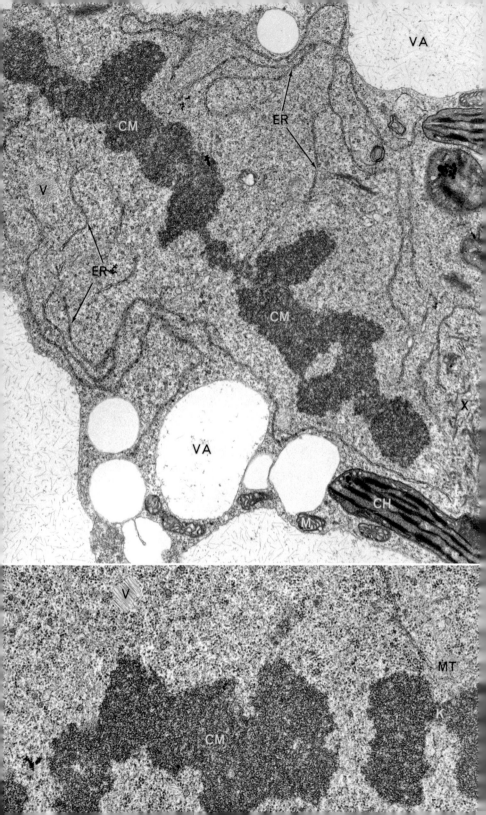

Fig. 41. Tobacco mosaic virus in relation to dividing nucleus in parenchyma cell of *Nicotiana tabacum*. Metaphase plate of chromosomes in center of cell as seen from a pole. One chromosome is in contact with virus aggregate. Vacuoles at both ends of the cell. × 9,000.

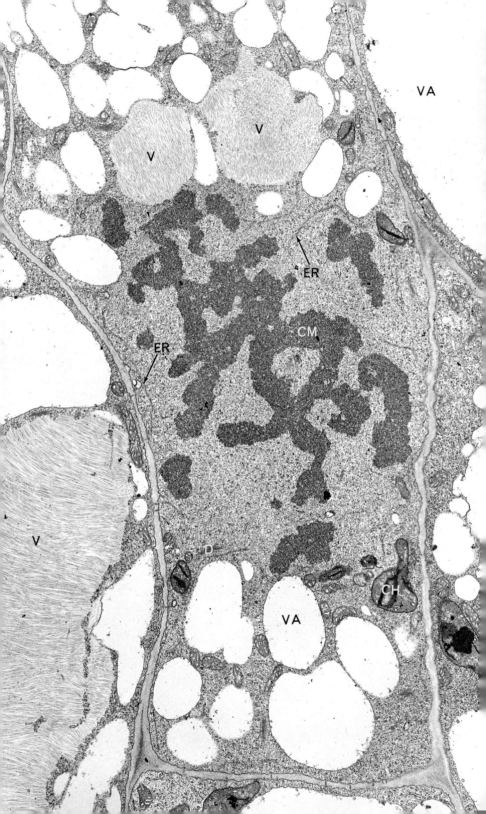

Figs. 42–45. Tobacco mosaic virus in relation to plastids in various cells of *Nicotiana tabacum*.

Fig. 42 (*above*). Two aggregates of virus in chloroplast in mesophyll cell. Lower aggregate has displaced grana, upper aggregate has formed a bulge in chloroplast envelope. × 30,000.

Fig. 43 (*middle, left*). Virus aggregate in young plastid in parenchyma cell. × 35,000.

Fig. 44 (*below, left*). Virus aggregates in plastid and cytoplasm of young tracheary element. × 30,000.

Fig. 45 (*below, right*). Virus aggregates in plastid and cytoplasm of young crystalliferous cell. × 28,000.

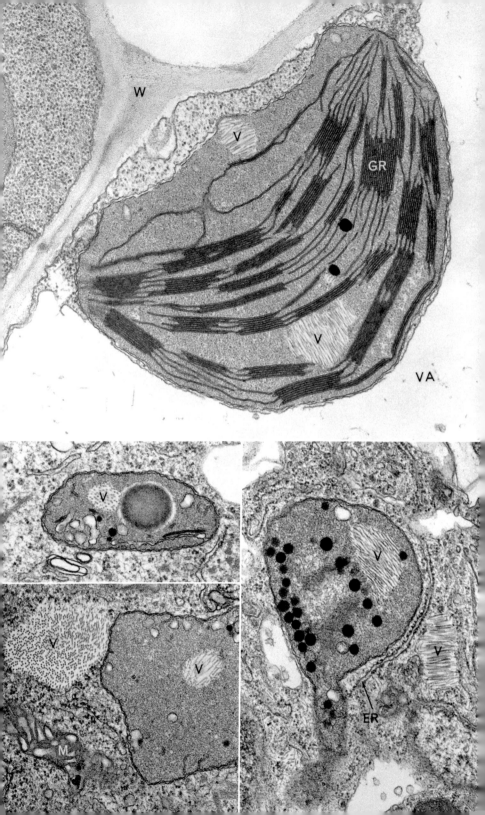

Figs. 46–49. Tobacco mosaic virus in relation to sieve-element plastids in *Nicotiana tabacum*.

Fig. 46 (*above*). Plastids and P-protein in young sieve element free of virus particles. × 41,000.

Fig. 47 (*middle, left*). Plastids in mature sieve element free of virus particles. × 49,000.

Fig. 48 (*middle, right*). Plastid with virus aggregate in a young sieve element. × 45,000.

Fig. 49 (*below*). Plastids in mature sieve element, the median with several virus particles. × 49,000.

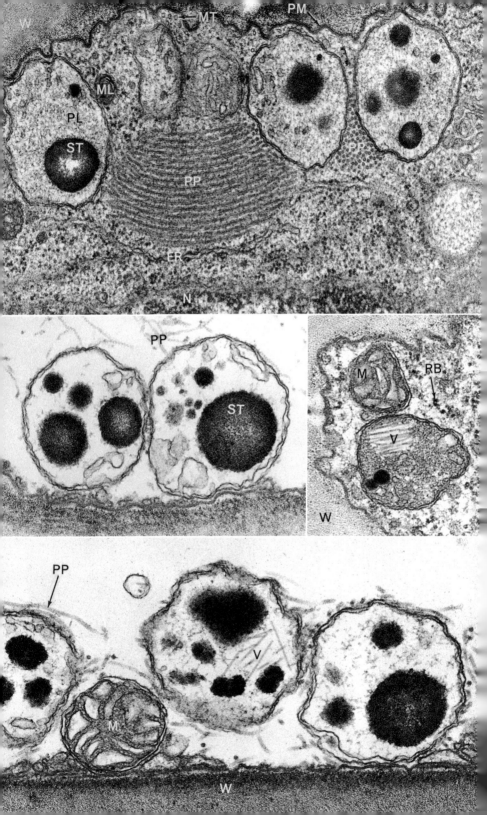

Figs. 50, 51. Tobacco mosaic virus in relation to cell types in *Nicotiana tabacum*.

Fig. 50 (*above*). Virus aggregates in guard cells of mature stoma. Bias cut has exposed nucleus in one guard cell only. × 6,250.

Fig. 51 (*below*). Virus-free guard cell of immature stoma with plasmodesmata (*arrows*) in walls between guard cell and adjacent epidermal cells. × 7,500.

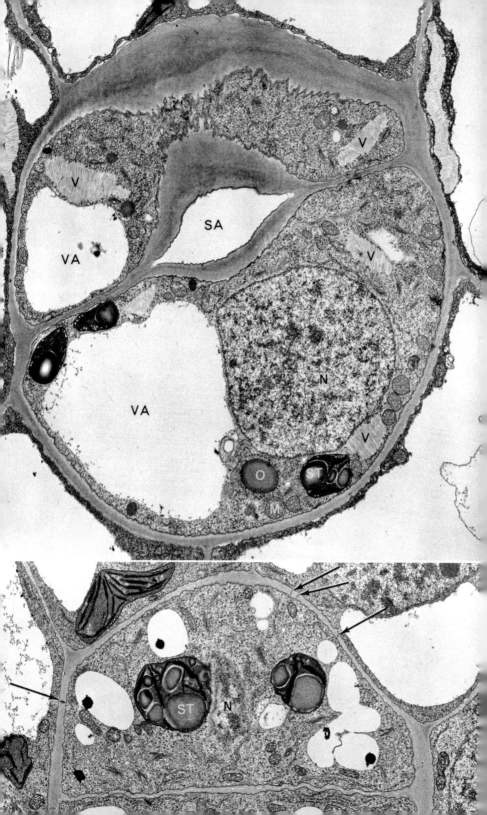

Fig. 52. Tobacco mosaic virus in relation to cell types in *Nicotiana tabacum*. Virus aggregate in a crystalliferous cell. Endoplasmic-reticulum cisternae in parallel arrangement next to virus aggregate. Crystalliferous cavities are devoid of crystals; these were probably dissolved in preparation. × 15,000.

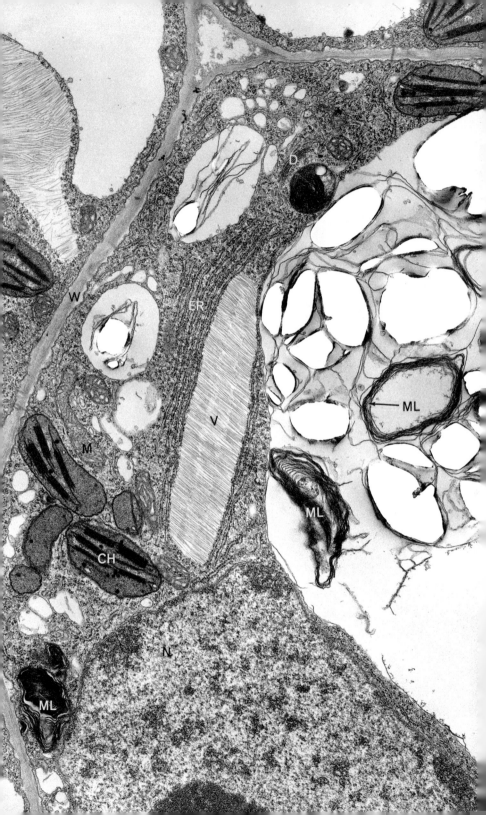

Figs. 53–55. Tobacco mosaic virus in relation to sieve element in *Nicotiana tabacum*.

Fig. 53 (*above, left*). Virus aggregate with particles in two opposite orientations near wall in a mature sieve element. Cisternae of endoplasmic reticulum form stack at left and one cisterna partly envelops aggregate. In *ER*, wider cisternae and narrower intercisternal spaces are filled with stained material. × 32,500.

Fig. 54 (*above, right*). Virus aggregate associated with endoplasmic reticulum in a mature sieve element. *ER* cisternae and narrow intercisternal spaces are filled with stained material. × 32,500.

Fig. 55 (*below*). Virus rods in various orientations fill lumen of sieve element in advanced stage of differentiation. Some ribosomes occur among virus rods. × 26,000.

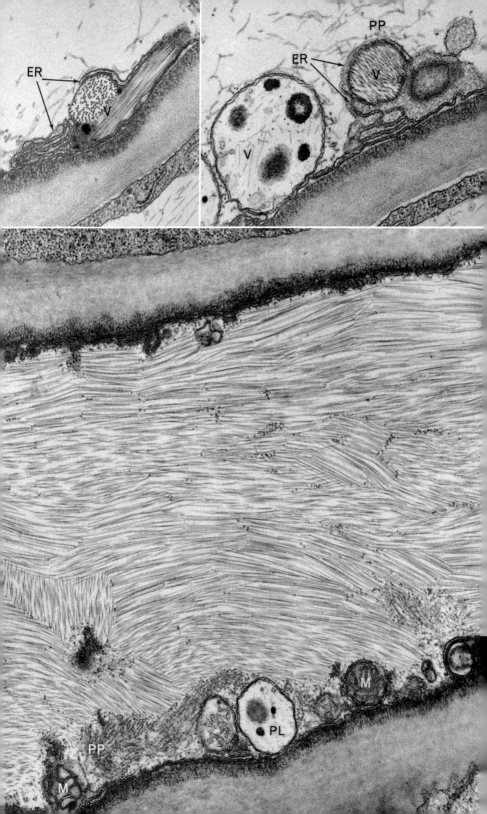

Fig. 56. Tobacco mosaic virus in relation to sieve element in *Nicotiana tabacum*. Virus, with particles in linear aggregations, fills lumen of mature sieve element. Virus rods are oriented uniformly longitudinally in the cell. Companion cell to the right is free of virus. Parenchyma cell in upper right corner contains virus particles in vacuole. × 24,000.

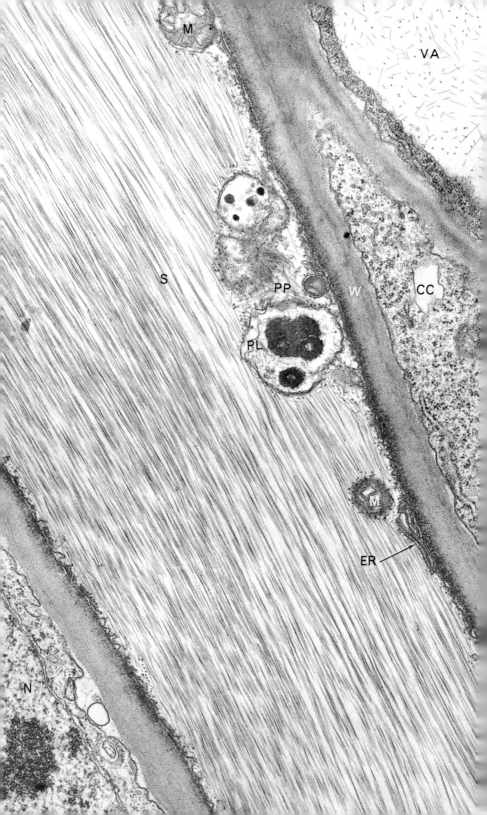

Figs. 57, 58. Tobacco mosaic virus in relation to sieve element in *Nicotiana tabacum*.

Fig. 57 (*above*). Virus aggregate in young sieve element. P-protein body to the right indicates early stage of differentiation of sieve element. Tonoplast-delimited vacuole to the left. × 15,000.

Fig. 58 (*below*). Groups of virus rods in young sieve element. P-protein partly dispersed, an indication of later stage of differentiation than is show in Figure 57. × 32,500.

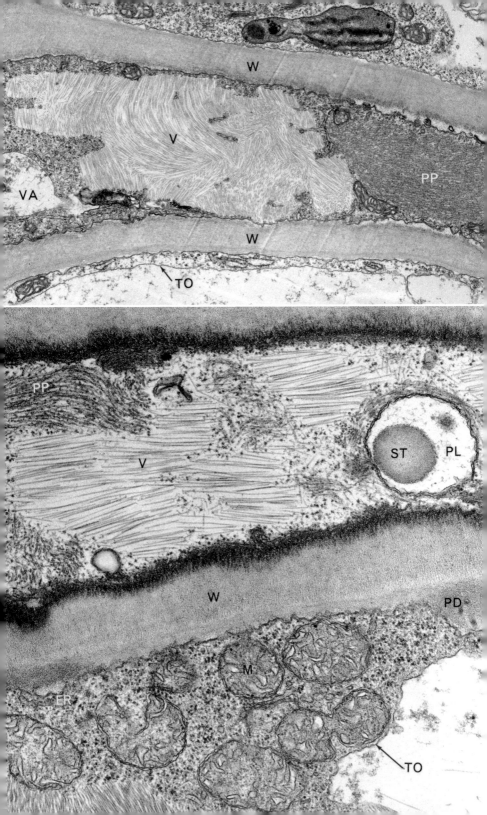

Fig. 59. Tobacco mosaic virus in relation to sieve element in *Nicotiana tabacum*. Portions of two young sieve elements with immature sieve plate between them. Upper sieve element contains an aggregate of virus, an X-body, and a P-protein body. The X-material is represented by bands composed of groups of three tubules. One pore site with a plasmodesma is visible in sieve plate. × 26,000.

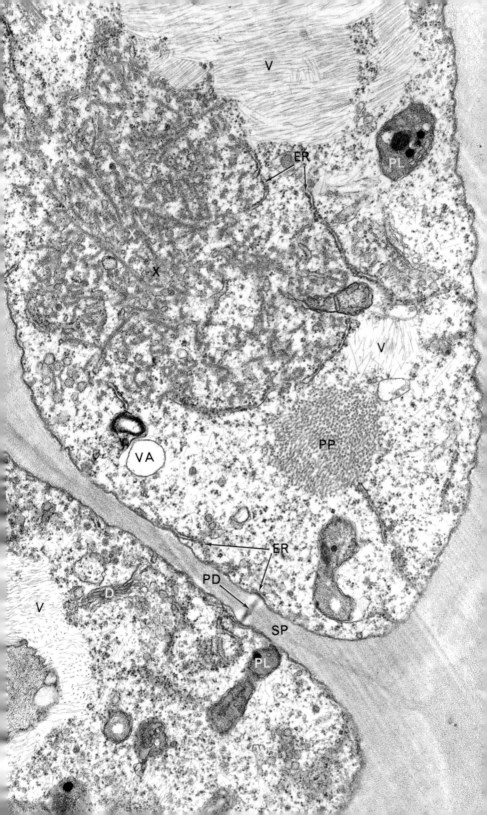

Fig. 60. Tobacco mosaic virus in relation to tracheary elements in *Nicotiana tabacum*. Portions of two mature tracheary elements with a wall bearing helical secondary thickenings (seen in section) between them. The primary wall component is largely disintegrated. Numerous virus rods in lumina of elements, in contrasting orientations in the two cells. × 30,000.

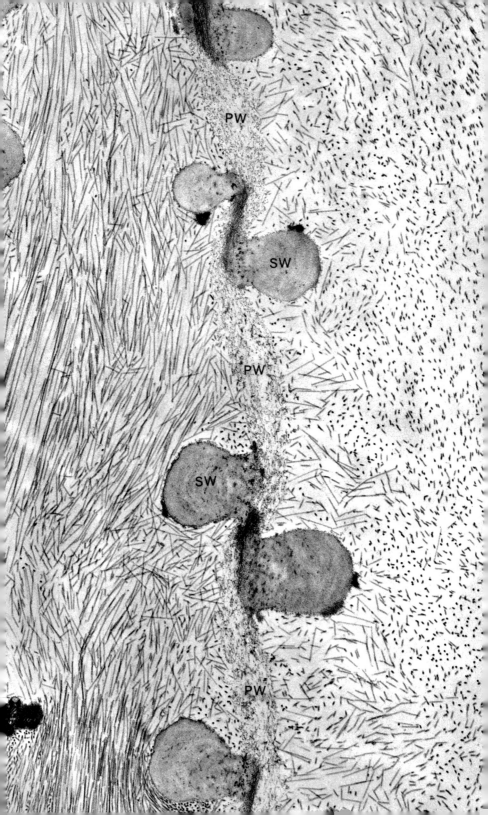

Figs. 61, 62. Tobacco mosaic virus in relation to tracheary elements in *Nicotiana tabacum*.

Fig. 61 (*above*). Portion of immature tracheary element with helical secondary wall (seen in section). Numerous small aggregates of virus particles are interspersed with ribosomes. × 31,000.

Fig. 62 (*below*). Portion of immature tracheary element with helical secondary wall (seen in section). X-material occurs between wall and nucleus. Somewhat inflated endoplasmic-reticulum cisternae with stained contents are mixed with X-material. Nucleus contains small aggregates of virus particles. × 31,000.

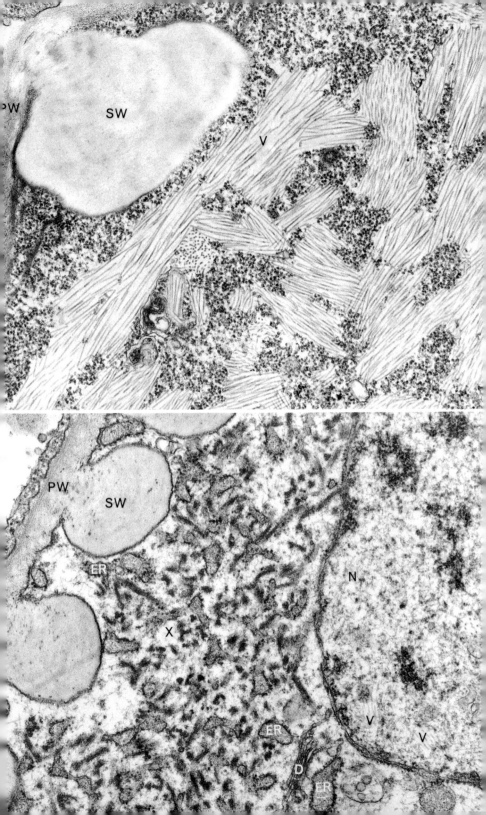

Fig. 63. Beet yellows virus in relation to sieve element in *Beta vulgaris*. Virus particles are randomly distributed in lumen of mature sieve element. Portion of normal-appearing plastid at *PL*. Identity of membrane-limited vesicles with sparse contents is uncertain. Parenchyma cell to the right contains large amount of virus in aggregates interspersed with ribosomes and mitochondria. × 30,000.

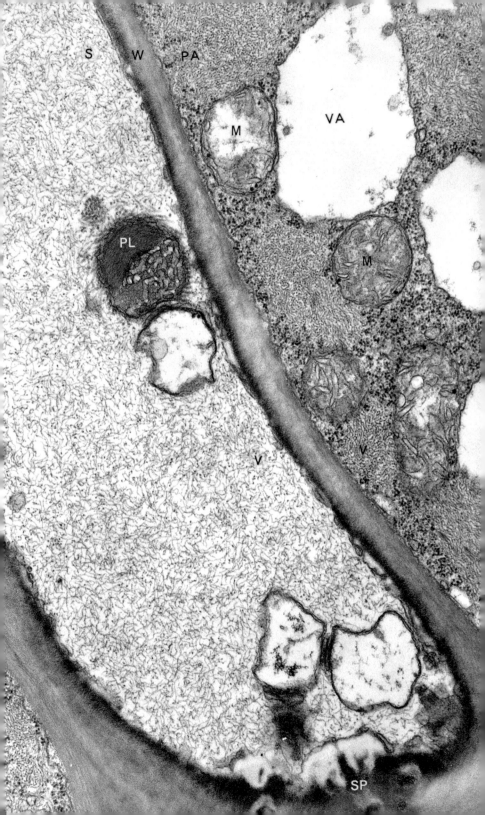

Figs. 64–66. Structure of plasmodesmata.

Fig. 64 (*above*). Branched plasmodesmata in a wall between two parenchyma cells from healthy leaf of *Cucurbita maxima*. In this material, fixed with potassium permanganate, profiles of endoplasmic-reticulum cisternae are particularly sharp. Association of cisternae with plasmodesmata is indicated. × 50,000.

Fig. 65 (*below, left*). Plasmodesmata in wall between two parenchyma cells in noninfected *Nicotiana tabacum*. The plasmodesmata have median cavities. At arrow, the *ER* membrane nearest the wall is bent toward plasmodesmatal core. × 45,000.

Fig. 66 (*below, right*). Plasmodesma in a wall between young sieve element (*upper cell*) and parenchyma cell (*lower cell*) in *Nicotiana tabacum*. Infected with TMV but cells free of virus. Plasmodesma contains a central core. Plasmalemmas on two surfaces of wall are continuous with lining of plasmodesmatal canal. Callose surrounds canal on the side of sieve element. On opposite side, plasmodesma is branched. Loosely structured wall layer under plasmalemma on sieve-element side. × 49,000.

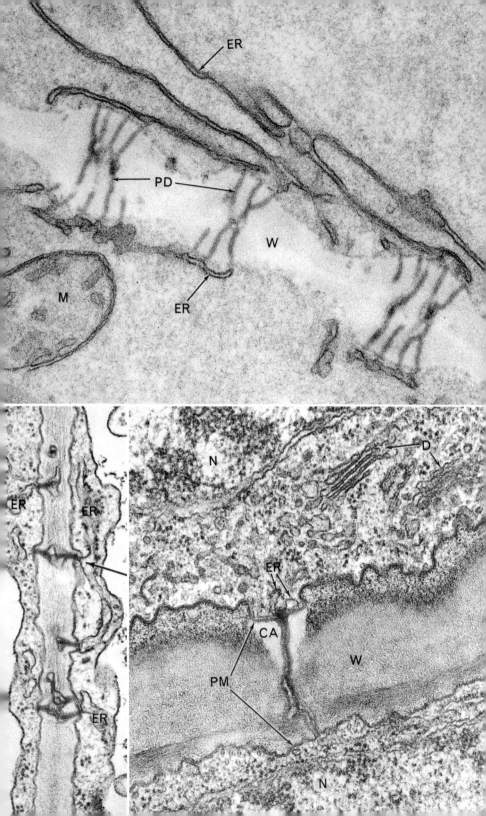

Figs. 67–70. Structure of plasmodesmata.

Fig. 67 (*above, left*). Group of plasmodesmata in transection seen in parenchyma-cell wall in noninfected leaf of *Beta vulgaris*. Each plasmodesma is lined with plasmalemma and has a tubular core. × 48,000.

Fig. 68 (*above, right*). Partial longitudinal section of branched plasmodesma in wall between sieve element (not shown) and parenchyma cell in noninfected leaf of *Nicotiana tabacum*. To the right, tubular core is connected to a cisterna of endoplasmic reticulum. × 60,000.

Fig. 69 (*middle*). Longitudinal section of plasmodesmata in wall between sieve element (*S*) and parenchyma cell in healthy *Cucurbita maxima* leaf. Plasmodesmata are branched on side of parenchyma cell. × 32,500.

Fig. 70 (*below*). Plasmodesmata in face view in oblique section through wall between sieve element (*below, left*) and parenchyma cell (*above*) in healthy *Cucurbita maxima* leaf. Groups of pores indicate groups of branches of single plasmodesmata. Single pores on sieve-element side (*arrows*). × 26,000.

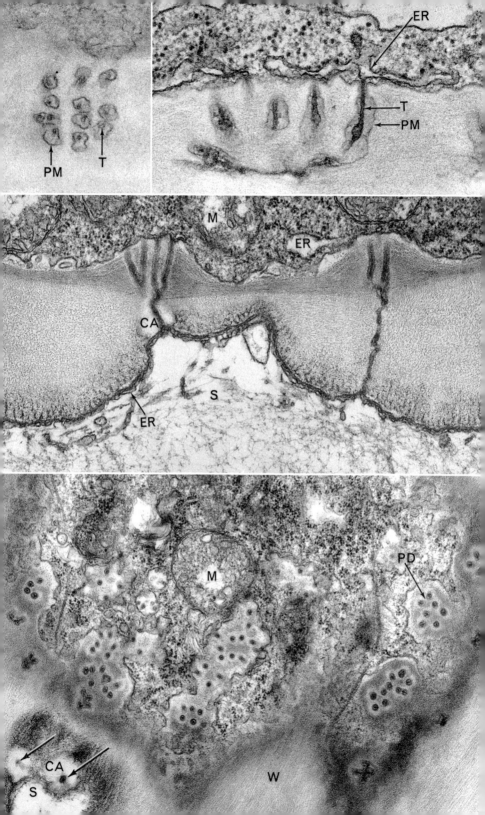

Figs. 71, 72. Structure of sieve plate in sieve element of *Nicotiana tabacum*. Plant infected with TMV but cells free of virus.

Fig. 71 (*large picture*). Section of portion of sieve plate. Three pores lined with callose and filled with tubules of P-protein are in view. Accumulation of P-protein (slime plug) occurs on one side of plate. × 56,000.

Fig. 72 (*above, insert*). Enlarged view of P-protein from Figure 71. Transverse sections of tubules show electron-lucent centers. × 120,000.

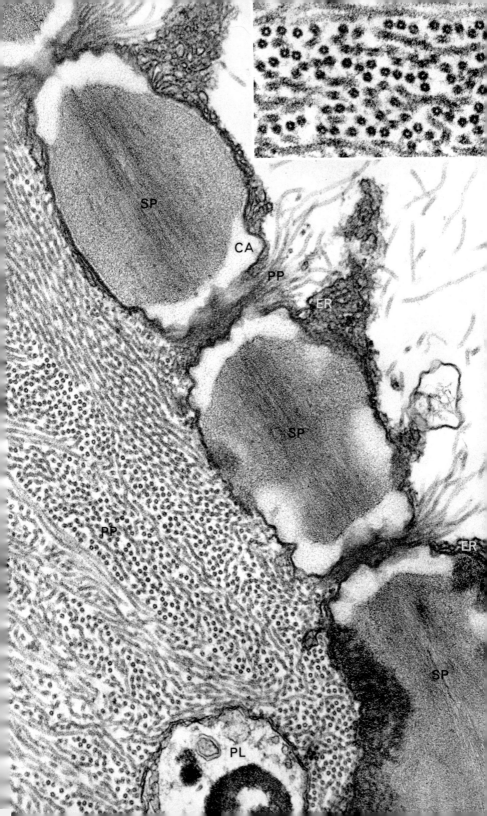

Figs. 73–79. Beet yellows virus in relation to various pores in walls of phloem in leaf of *Beta vulgaris*.

Fig. 73 (*above, left*). Sieve plate with open pores in sieve tube from noninfected plant. × 12,000.

Fig. 74 (*above, center*). Virus particles in plasmodesma canal in wall between two parenchyma cells. × 80,000.

Fig. 75 (*above, right*). Virus particles in transection of plasmodesma on sieve-element side in wall between sieve element and parenchyma cell. × 80,000.

Figs. 76 (*middle, left*) and 77 (*middle, center*). Virus particles in branched plasmodesma in wall between sieve element (*above*) and parenchyma cell. Both, × 60,000.

Fig. 78 (*middle, right*). Virus particles in transections of plasmodesmatal branches in wall between sieve element and parenchyma cell. × 60,000.

Fig. 79 (*below*). Virus particles and P-protein in pores of sieve plate. × 30,000.

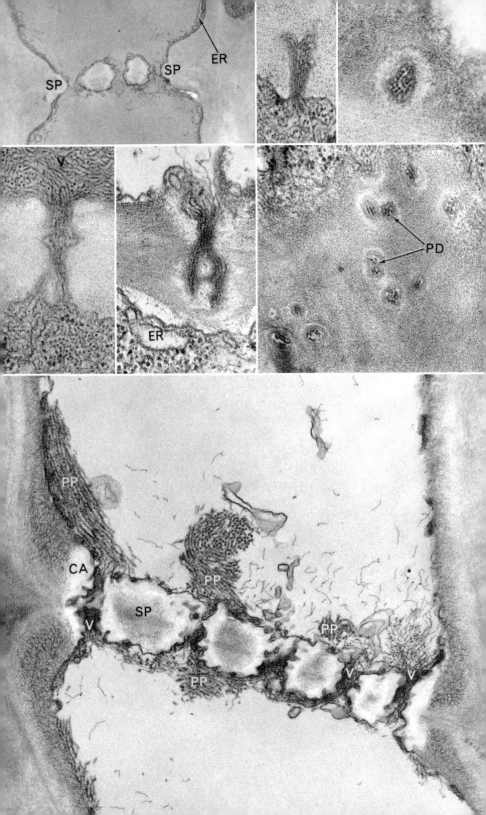

Fig. 80. Beet yellows virus in relation to sieve plate in sieve tube from leaf vein of *Beta vulgaris*. Three pores lined with callose are in view. One is cut through middle and is shown filled with virus particles. P-protein has plugged another pore. Some P-protein tubules are extended toward virus-containing pore (*arrow*). × 45,000.

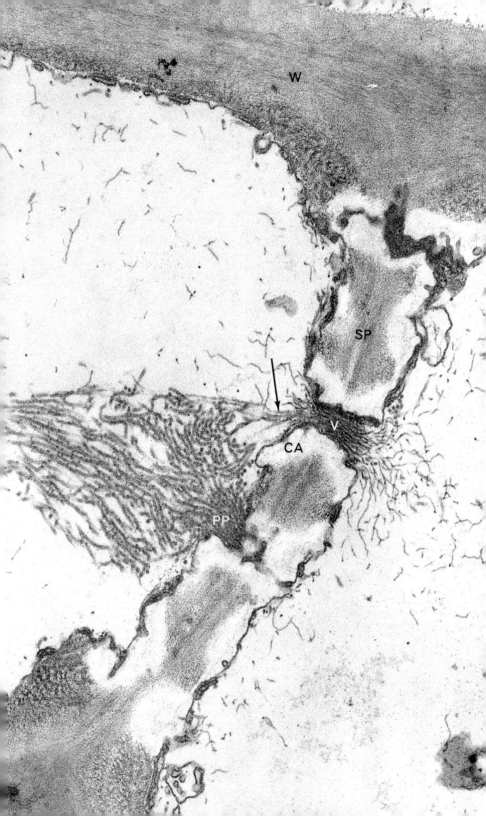

Fig. 81. Tobacco mosaic virus in dividing cells of *Nicotiana tabacum*. Two young mesophyll cells derived from a recently divided precursory cell. Each shows a nucleus, chloroplasts, vacuoles, and one large aggregate of virus in cytoplasm. × 6,000.

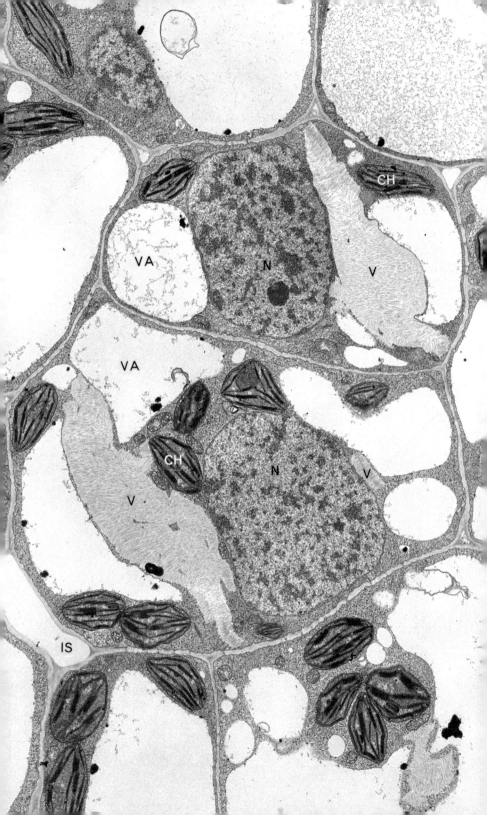

Fig. 82. Tobacco mosaic virus in dividing cells of *Nicotiana tabacum*. Highly vacuolated mesophyll cell in mitosis. Metaphase chromosomes occur in cytoplasmic layer, the phragmosome. Virus aggregate occurs between chromosomes and cell wall. × 7,500.

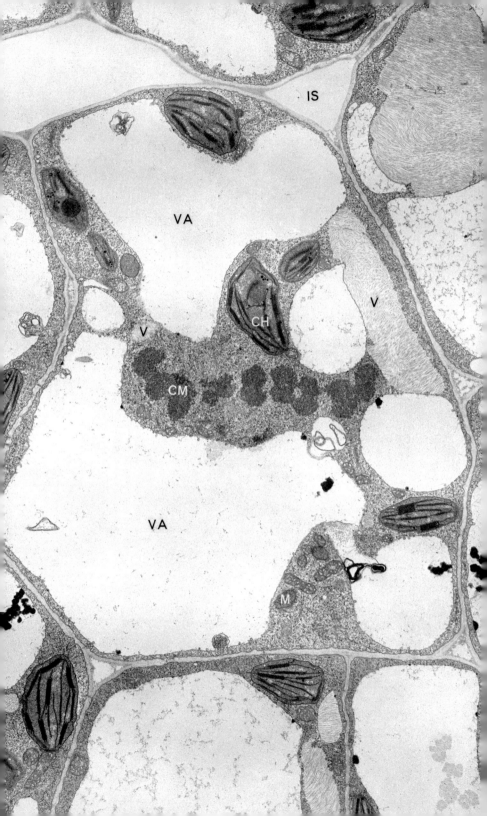

Figs. 83, 84. Tobacco mosaic virus in dividing cells of *Nicotiana tabacum*.

Fig. 83 (*above*). Dividing cell in stage of cytokinesis. Daughter nuclei have formed envelopes. Incomplete cell plate occurs between nuclei. Cell plate has not yet become united with mother-cell walls (*arrows*). Virus aggregates occur in both daughter cells. × 9,000.

Fig. 84 (*below*). Portion of dividing cell in stage of cytokinesis. Virus aggregate occurs between mother-cell wall and growing cell plate. One vesicle of cell plate almost touches virus aggregate (*arrow*). × 20,000.

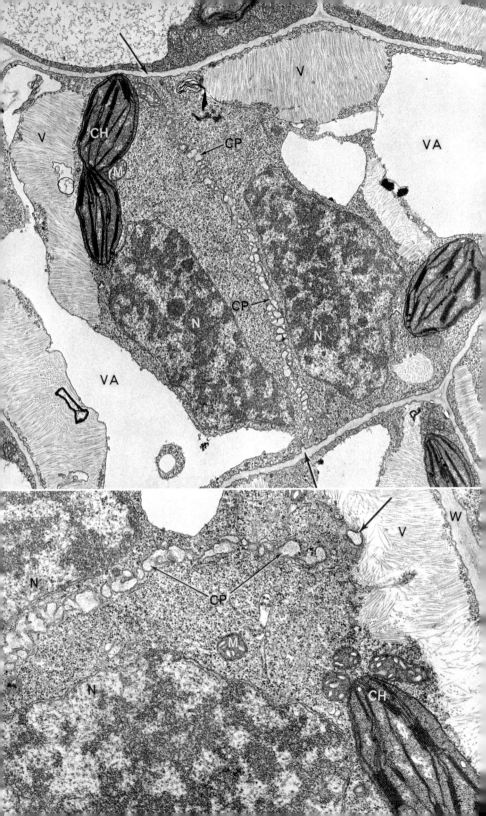

Figs. 85, 86. Tobacco mosaic virus in dividing cells of *Nicotiana tabacum*.

Fig. 85 (*above*). Portion of dividing cell in stage of cytokinesis. Virus aggregate located between mother-cell wall and cell plate appears to be partly disrupted. × 20,000.

Fig. 86 (*below*). Portion of dividing cell in stage of cytokinesis. Cell plate has nearly reached mother-cell wall. Virus aggregates occur on both sides of cell plate close to mother-cell wall. × 20,000.

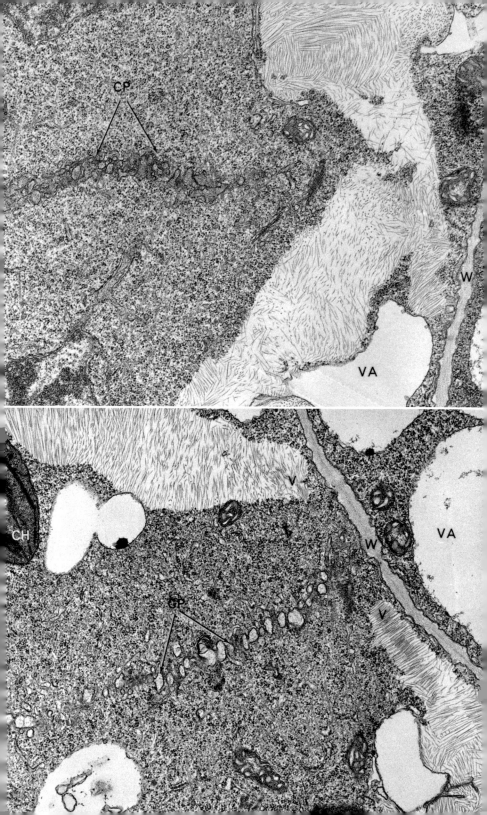

Figs. 87, 88. Tobacco mosaic virus in dividing cells of *Nicotiana tabacum*.

Fig. 87 (*above*). Portion of dividing cell in stage of cytokinesis. Cell plate has reached mother-cell wall but component vesicles are still sparse and microtubules of phragmoplast are evident on both sides of latest-formed part of cell plate. × 30,000.

Fig. 88 (*below*). Portion of dividing cell in stage of cytokinesis. Cell-plate vesicles occur close to the mother-cell wall. Microtubules of the phragmoplast are evident above the cell plate but are lacking below it where some X-material is aggregated. Triplets of X-tubules at arrows. × 45,000.

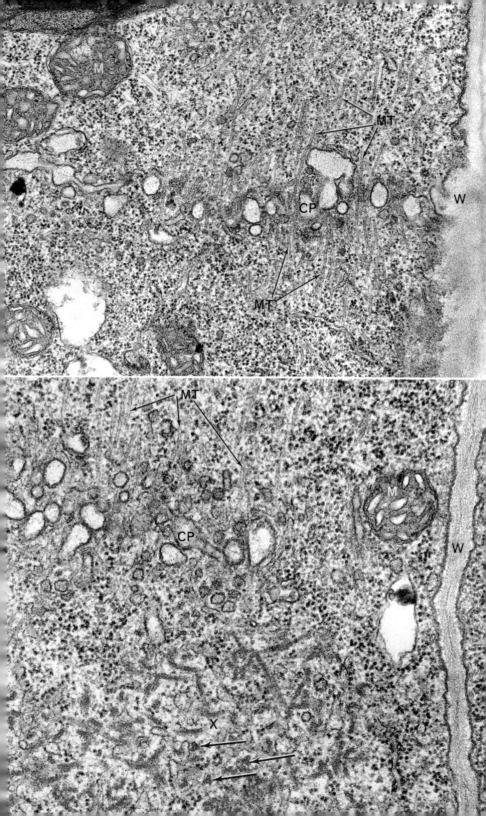

Figs. 89, 90. Chloroplasts from noninfected leaf of *Beta vulgaris*.

Fig. 89 (*left*). Thylakoids are organized into grana and intergrana lamellae. Small osmiophilic globules in stroma. Microbody next to plastid above. Thylakoid system is somewhat displaced toward side away from cell wall (not shown). × 23,000.

Fig. 90 (*right*). Thylakoid system as in Figure 89 but not displaced; osmiophilic globules somewhat larger and less electron opaque. × 26,000.

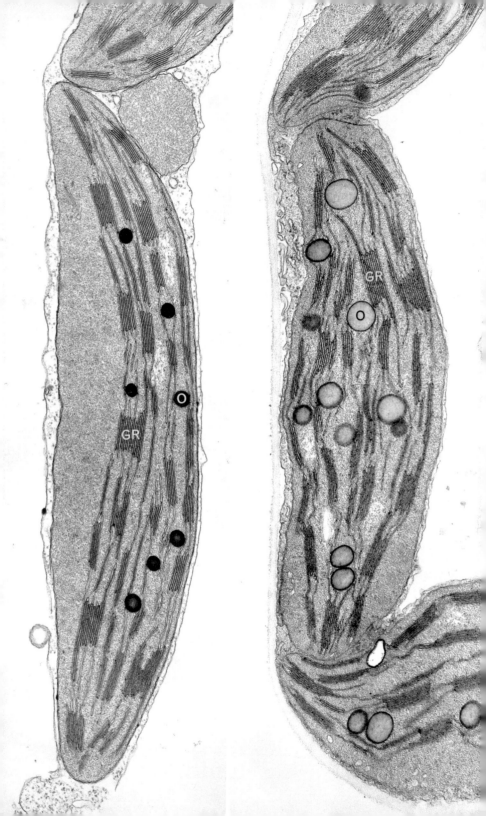

Figs. 91–93. Chloroplasts of *Beta vulgaris* from noninfected leaf (Fig. 91) and leaves infected with beet yellows virus (Figs. 92 and 93).

Fig. 91 (*above, left*). Osmiophilic material not in globules but as irregular bodies because of fixation with potassium permanganate. Thylakoid system somewhat displaced toward side away from cell wall. × 36,000.

Fig. 92 (*above, right*). Thylakoid system deranged at *MB* into uniform aggregate of membranes forming entities with short profiles. Cell contains no virus. × 30,000.

Fig. 93 (*below*). Grana oriented transversely to long axis of plastid. Intergrana lamellae are sparse. Plastid to the left has invagination pocket enclosing mitochondrion. Cell contains virus. × 26,000.

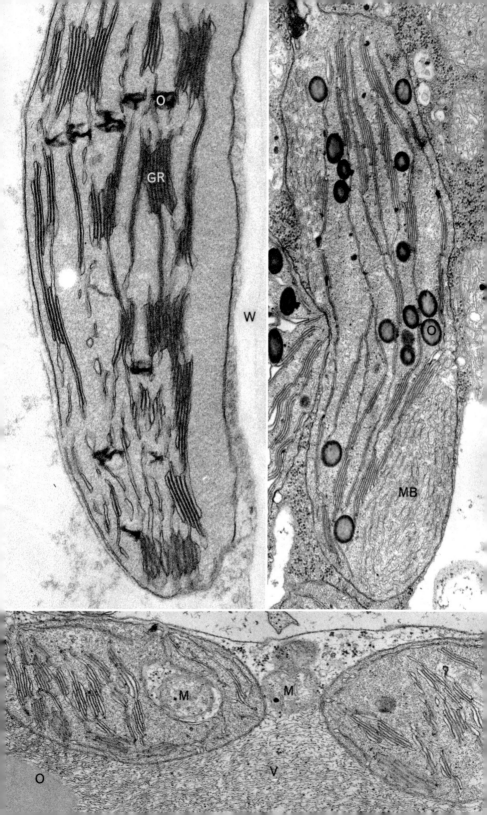

Figs. 94–96. Chloroplasts in leaves of *Beta vulgaris* infected with beet yellows virus. Cells contain virus particles.

Fig. 94 (*above*). Lamellae in plastid with long profiles not organized into grana-intergrana lamellae system. Stroma is dense and contains osmiophilic globules, unidentified electron-opaque material (*arrow*), and vesicles near plastid periphery (one apparently extruded). Plastid is lobed at lower end. Flaky electron-opaque material is associated with ribosomes to the left of plastid. × 32,500.

Fig. 95 (*middle*). Chloroplast similar to that in Figure 94. Vesicle (*VE*) indicates its origin from inner layer of plastid envelope. Vesicles occur in the cytoplasm to the right of plastid, some in groups surrounded by a common membrane. × 32,500.

Fig. 96 (*below*). Strongly modified chloroplast with dense stroma and osmiophilic globules in center. Lamellae with long profiles are displaced toward plastid periphery. × 30,000.

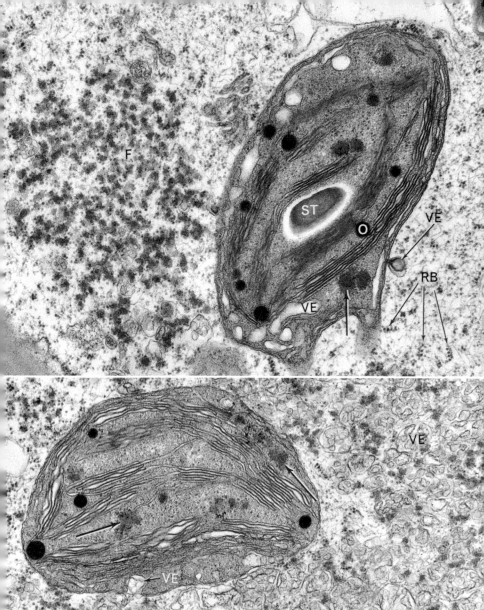
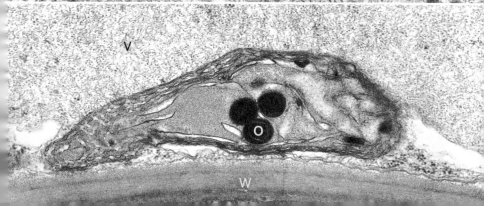

Figs. 97–99. Chloroplasts in leaves of *Beta vulgaris* infected with beet yellows virus. Cells contain no virus.

Fig. 97 (*above*). Thylakoid system consists mainly of grana which are composed of unusually large numbers of thylakoids. Stroma contains large osmiophilic globules. × 30,000.

Fig. 98 (*middle*). Chloroplast as in Figure 97 but also contains starch grains. × 30,000.

Fig. 99 (*below*). Two thylakoid systems enclosed in common stroma and single envelope. Each includes osmiophilic globules and starch. × 22,500.

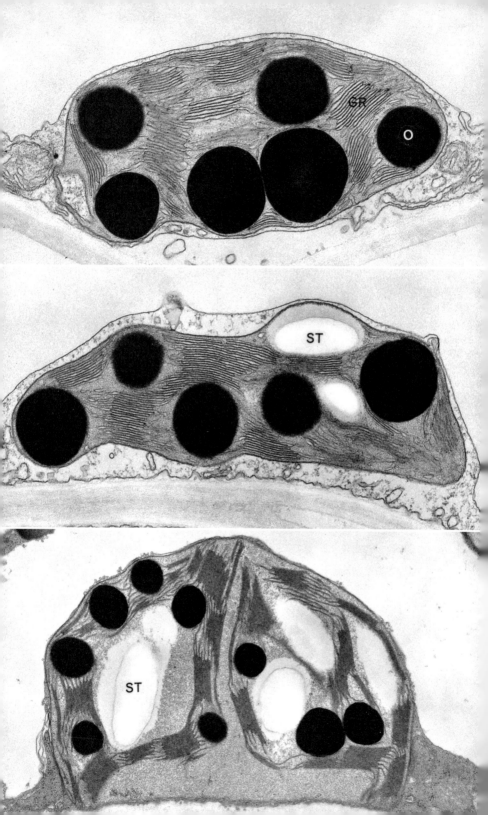

Figs. 100–105. Variations in structure of osmiophilic globules in cytoplasm (Fig. 100) and chloroplasts (Figs. 101–105) of leaves of *Beta vulgaris* infected with beet yellows virus. The cells contain no virus.

Fig. 100 (*above, left*). Smooth globules are free in parietal cytoplasm. × 30,000.

Fig. 101 (*above, right*). Globules show differentiation of periphery into paracrystalline system of rods. Globule to the left cut through middle, the one to the right, tangentially. × 60,000.

Fig. 102 (*middle, left*). Paracrystalline rod system is shallow. × 60,000.

Fig. 103 (*middle, right*). Paracrystalline rod system is extensive. Amorphous core appears deeply eroded. × 60,000.

Fig. 104 (*below, left*). Small paracrystalline rod systems in isolated regions of globules. × 60,000.

Fig. 105 (*below, right*). Alveolate peripheral region surrounds amorphous core. ×45,000.

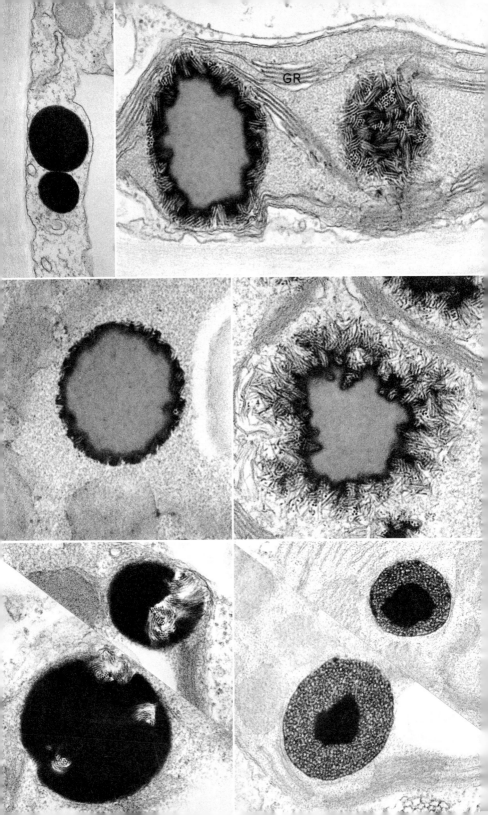

Figs. 106–109. Inclusions in cytoplasm (Fig. 106) and chloroplasts (Figs. 107–109) in leaves of *Beta vulgaris* infected with beet yellows virus. No virus particles in cells in Figures 107–109.

Fig. 106 (*above*). Portion of cell containing osmiophilic globules in cytoplasm which is packed with virus particles. At right, protrusions filled with virus particles have almost occluded a vacuole. × 25,000.

Fig. 107 (*below, left*). Osmiophilic globule and aggregate of phytoferritin as they appear in sections stained with uranyl acetate and lead citrate. × 80,000.

Fig. 108 (*below, middle*). Aggregate of phytoferritin as it appears in sections not stained with uranyl acetate and lead citrate. × 100,000.

Fig. 109 (*below, right*). Aggregate of phytoferritin from stained section, highly magnified and indicating, at arrows, resolution of granules into subunits. × 240,000.

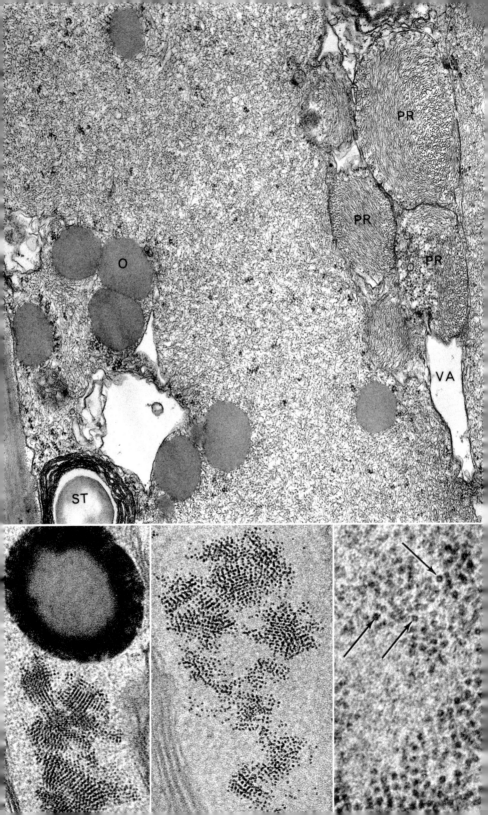

Figs. 110, 111. Chloroplasts of *Nicotiana tabacum*.

Fig. 110 (*left*). Chloroplast from noninfected leaf. Thylakoid system differentiated into grana and intergrana lamellae. Sparse osmiophilic material in stroma and at margins of grana thylakoids. Peripheral stroma contains some vesicles. × 40,000.

Fig. 111 (*right*). Chloroplast from leaf infected with tobacco mosaic virus. Cell devoid of virus particles. Thylakoids mostly of long profiles; no well-differentiated grana are present. Group of osmiophilic globules in center of stroma. × 40,000.

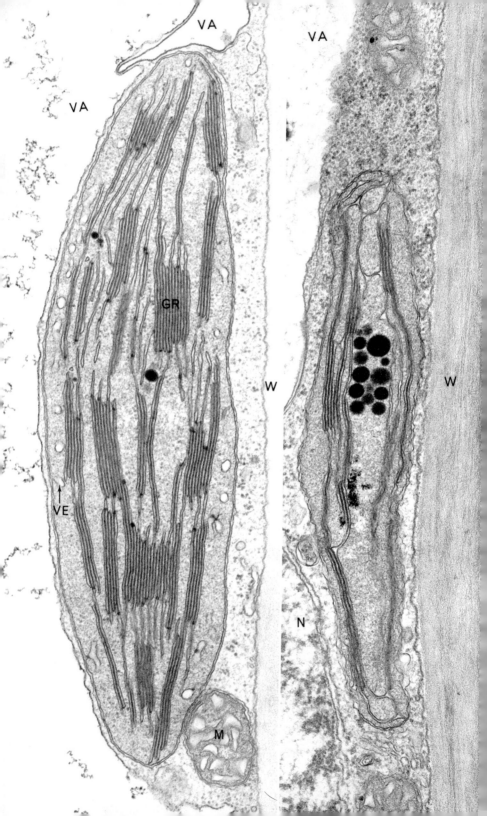

Figs. 112, 113. Chloroplasts from leaves of *Nicotiana tabacum* infected with tobacco mosaic virus. Cells contain virus particles.

Fig. 112 (*above*). Two thylakoid systems enclosed in common stroma and single envelope. Virus aggregate occurs in stroma. No other abnormalities are in evidence. × 30,000.

Fig. 113 (*below*). Thylakoid system is displaced toward side away from wall. Grana-free stroma region contains a network of membranes. Otherwise plastid is normal in structure. Microbody with crystal next to plastid above. × 24,000.

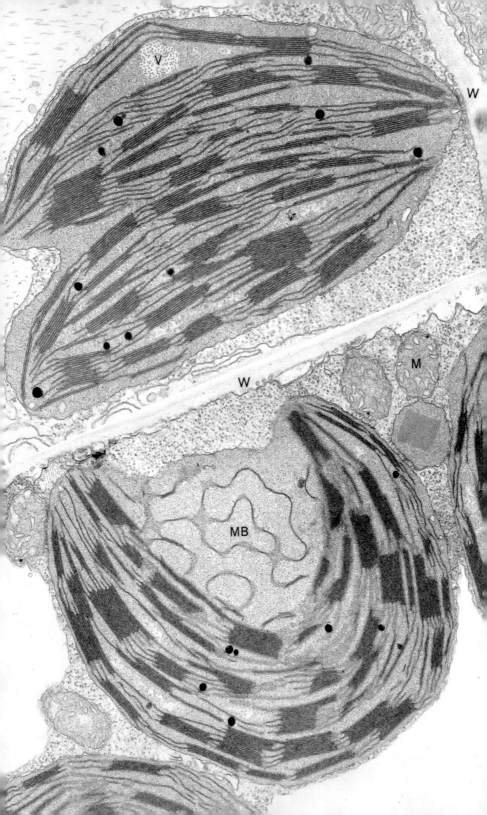

Figs. 114–116. Sieve-element plastids from noninfected leaves of *Beta vulgaris*.

Fig. 114 (*above*). In plastid to the right, fibrous ring, seen in face view, surrounds aggregate of inflated membranous entities. P-protein covers double-membraned plastid envelope. From mature cell. × 49,000.

Fig. 115 (*middle*). Sections of fibrous ring in plastid flank aggregate of membranous entities. P-protein tubules on surface of plastid cut transversely in region of fibrous ring. Unidentified electron-opaque material occurs among membranous aggregate. Partly dispersed P-protein to the left of plastid. From mature cell. × 49,000.

Fig. 116 (*below*). Plastid in young cell. Fibrous ring in sectional view. No P-protein on surface of plastid; it was not yet dispersed in cell. × 49,000.

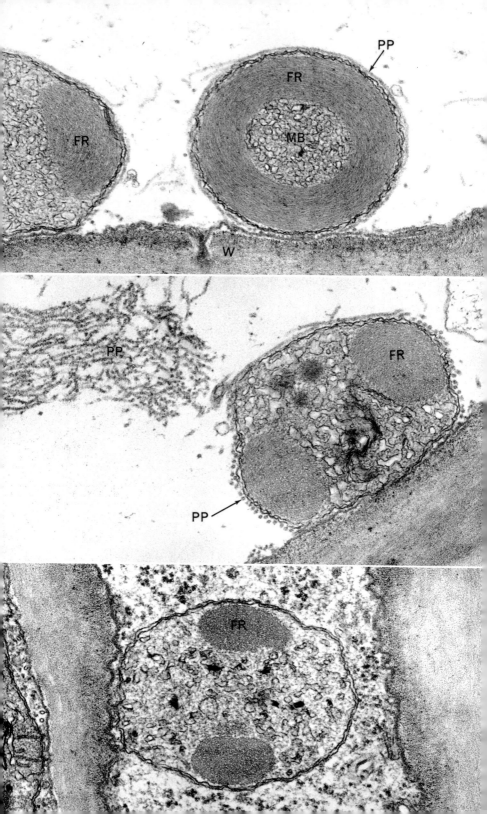

Figs. 117–120. Modified sieve-element plastids from leaves of *Beta vulgaris* infected with beet yellows virus. No virus particles are present in cells. Plastids are somewhat larger than normal.

Fig. 117 (*above*). Inner membranes form lamellae with long profiles. Some granalike stacks are present. Fibrous ring, seen in section, is thin. Large osmiophilic globule in center, P-protein on surface. × 32,500.

Fig. 118 (*middle, left*). Inner membrane system abnormal in form and distribution. Conspicuous osmiophilic globules in stroma. P-protein on surface. × 32,500.

Fig. 119 (*middle, right*). Stroma thin, inner membranes sparse. P-protein on surface. × 32,500.

Fig. 120 (*below*). Considerably degenerated plastids. Stroma thin, inner membranes sparse, fibrous ring dispersed or absent. P-protein on surface. × 32,500.

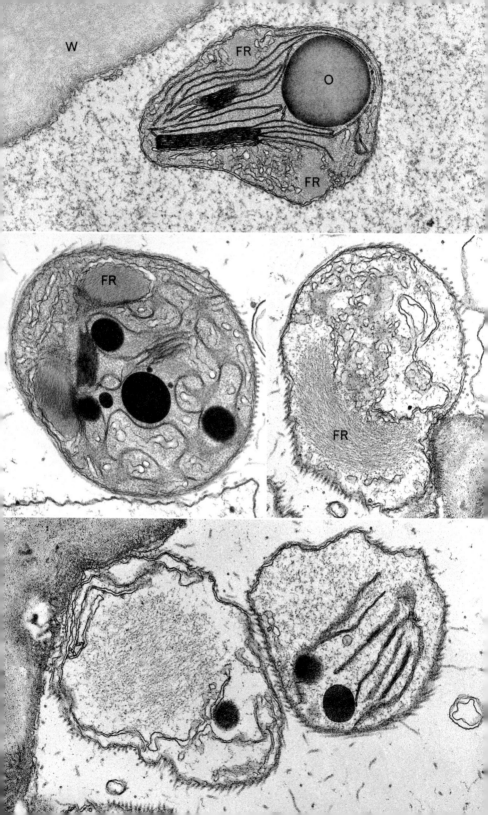

Fig. 121. Transverse section of phloem from a small vein of noninfected leaf of *Beta vulgaris*. The two sieve elements in view show characteristic thin layer of cytoplasm along wall. Associated parenchyma cells are vacuolated to various degrees. Each sieve element occupies the center in a group of cells, three of which (cells numbered *1, 2, 3*) indicate common origin with sieve element by their spatial relation to it. × 8,800.

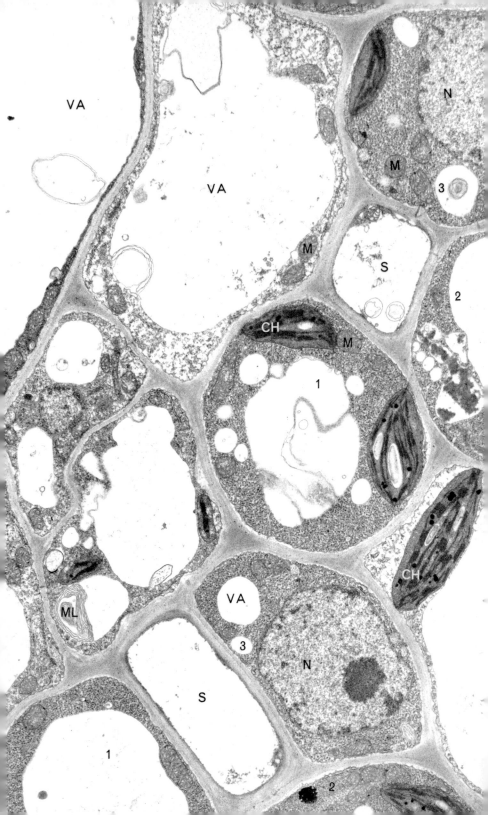

Fig. 122. Longitudinal section through immature sieve element and associated parenchyma cells. From phloem of small vein in a noninfected leaf of *Beta vulgaris*. In sieve element, vacuole is delimited by tonoplast, endoplasmic reticulum is in parietal position, and ribosomes are sparse. Rich complements of ribosomes in parenchyma cells (*left and below*). × 49,000.

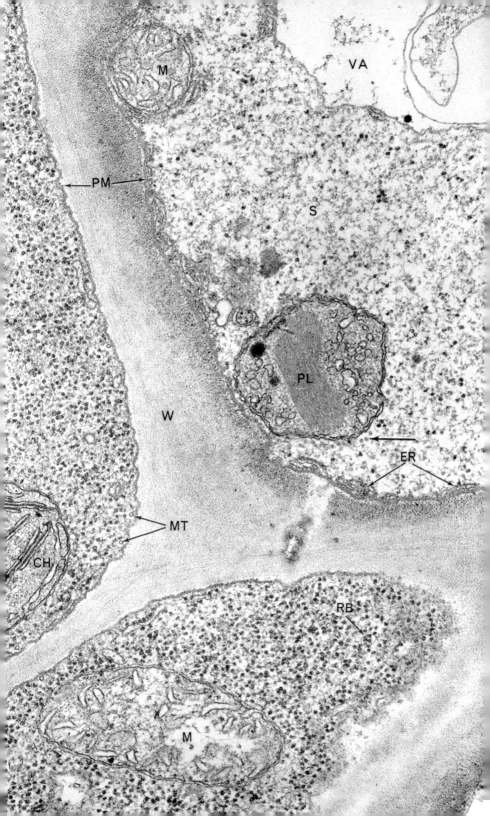

Fig. 123. Transverse section of phloem from a small vein of leaf of *Beta vulgaris* infected with beet yellows virus. The three sieve elements in view are associated with cells containing protoplasts of varying degrees of density. These cells are numbered *1–6*. Cell *1* is a sieve element completely filled with virus particles. Cells *2–5* are parenchyma cells containing masses of virus particles and degenerating protoplast components. Parenchyma cell *6* is free of virus. Its vacuole is large and chloroplasts contain starch. × 7,300.

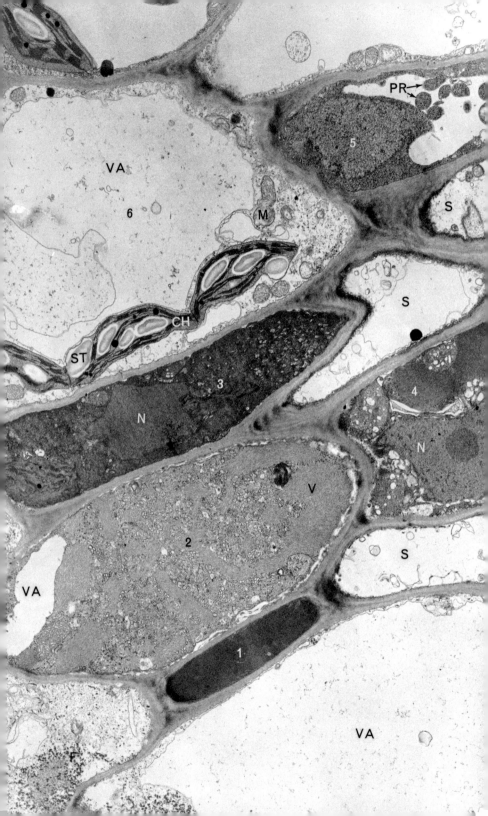

Fig. 124. Longitudinal section through phloem of small vein from leaf of *Beta vulgaris* infected with beet yellows virus. Parenchyma cell in center is densely filled with virus particles, some of which also occur in nucleus. Aggregate of vesicles appears above nucleus. Sieve element to the right is filled with virus particles. Parenchyma cells to the left are free of virus. × 7,500.

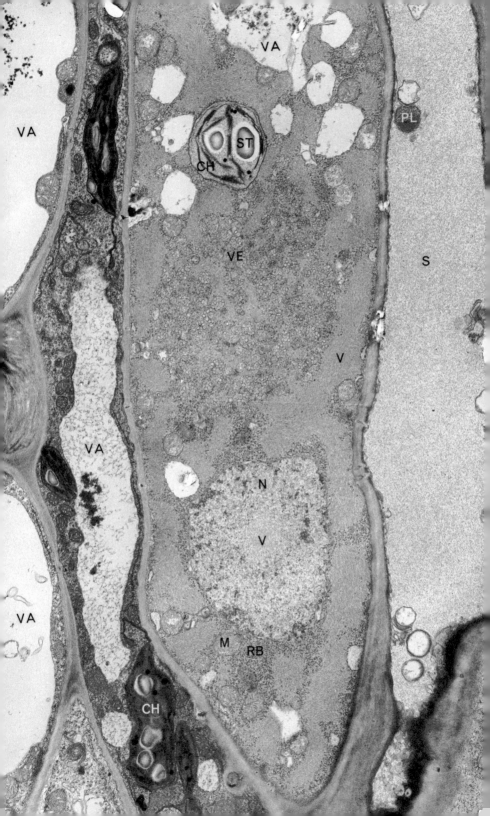

Fig. 125. Portion of phloem-parenchyma cell from small vein of leaf of *Beta vulgaris* infected with beet yellows virus. Vesicles in groups are interspersed with small aggregates of virus particles. Vesicles have conspicuous membranes and contain finely fibrous material often joined in a dot in the center. Above, flaky material associated with ribosomes. × 40,000.

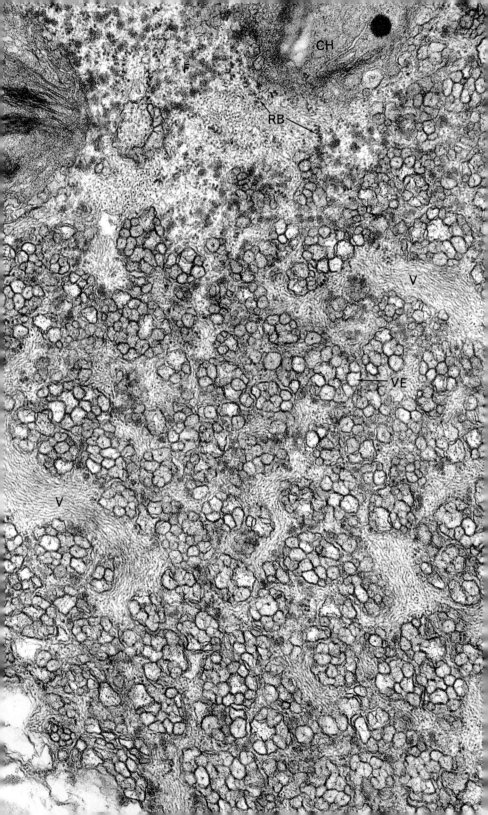

Fig. 126. Longitudinal section through phloem of small vein from leaf of *Beta vulgaris* infected with beet yellows virus. Parenchyma cell in center shows degenerated nucleus below and aggregate of vesicles above. Vesicles are associated with virus and flaky material. Virus, ribosomes, and flaky material also occur throughout cytoplasm. Chloroplasts are abnormal (compare with Figs. 94 and 95). Mitochondria appear to be normal. × 7,500.

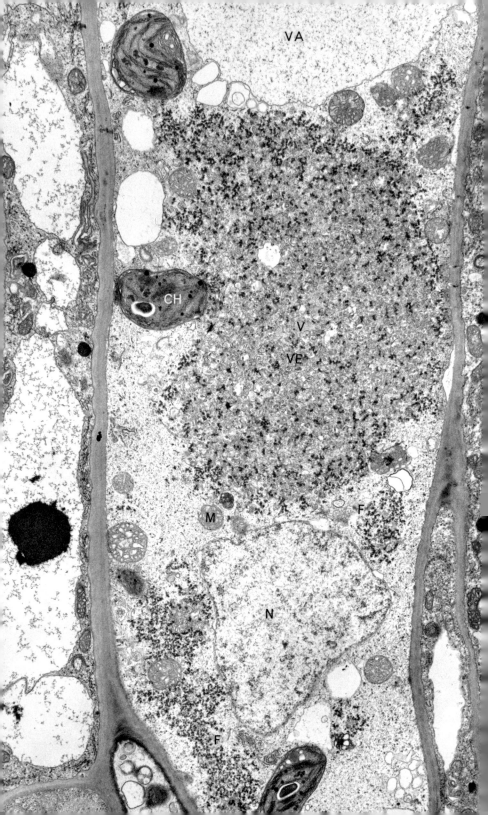

Fig. 127. Longitudinal section through severely degenerated phloem-parenchyma cells from small vein in leaf of *Beta vulgaris* infected with beet yellows virus. Normal-appearing virus particles are conspicuous. Their aggregates are embedded in partly necrosed contents of host cells. Among these are: broken-down chloroplasts, free starch grains, large aggregates of nearly empty vesicles, amorphous material highly opaque to electrons, and granules resembling ribosomes. Space above W is probably an artifact. × 9,000.

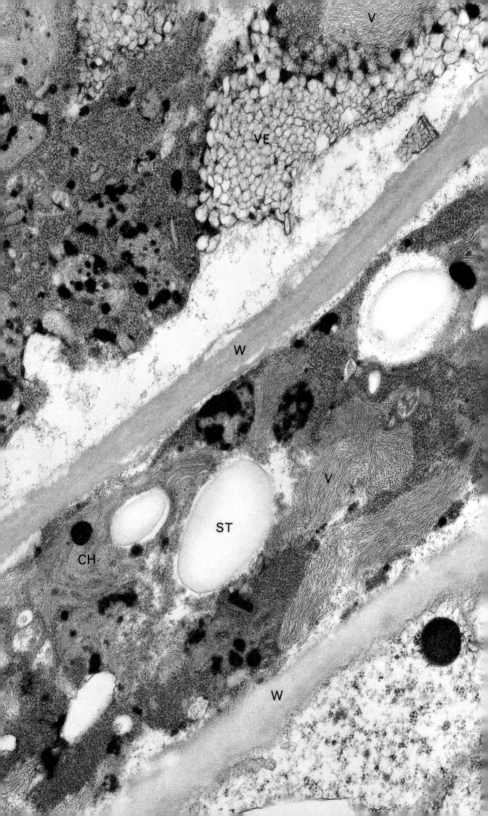

Fig. 128. Transection of mesophyll cell from leaf of *Nicotiana tabacum* infected with tobacco mosaic virus. Despite massive accumulation of virus and presence of X-body, the host protoplast, especially chloroplasts, appears to be normal. Vacuoles are free of virus. × 9,000.

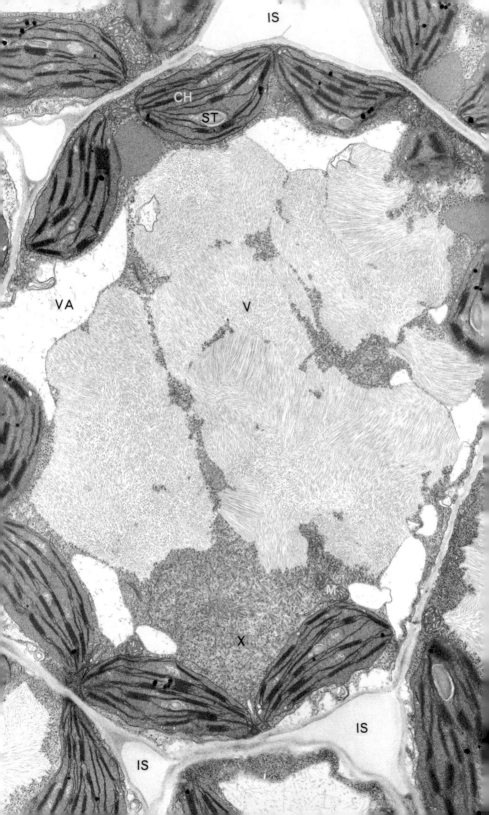

Fig. 129. Longitudinal section of phloem from small vein of leaf of *Nicotiana tabacum* infected with tobacco mosaic virus. No abnormalities in cells and no virus particles except in vacuole above to the right. Sieve element in center is immature. Portion of nucleus, vacuoles, and P-protein body are its dominant features. Companion cell to the right shows dense cytoplasm, many mitochondria, some dictyosomes, and two chloroplasts. The two parenchyma cells at margins of figure are highly vacuolated. × 9,500.

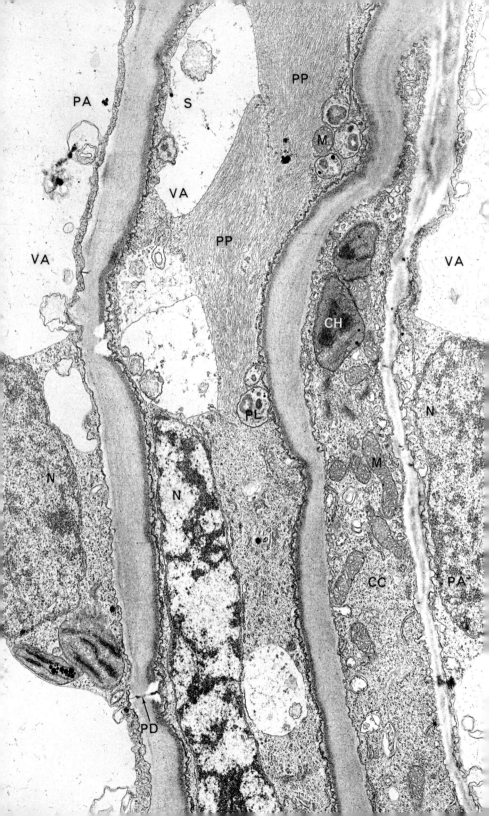

Fig. 130. Portion of phloem-parenchyma cell from small vein of leaf of *Nicotiana tabacum* infected with tobacco mosaic virus. Aggregates of virus particles and X-tubules appear side by side with aggregate of P-protein tubules of the host cell. The greater width of X-tubules as compared with P-protein tubules is discernible. (Compare with Figs. 20 and 72.) × 49,000.

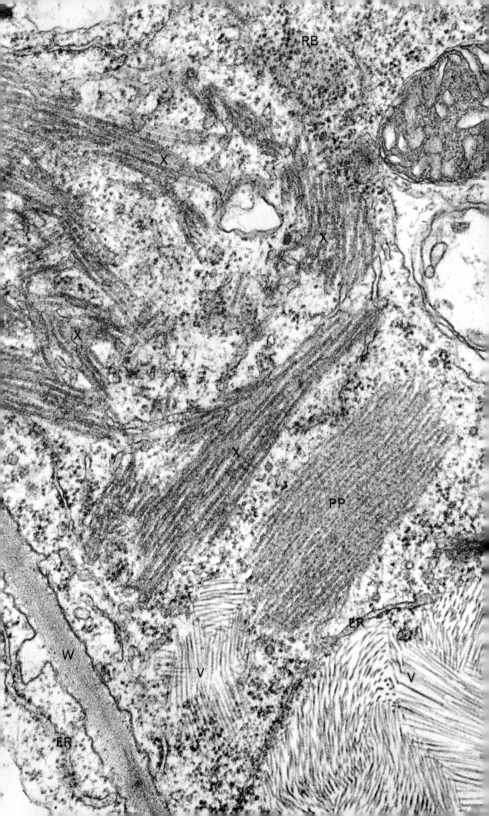

Figs. 131, 132. Portions of young sieve elements from small vein of leaf of *Nicotiana tabacum* infected with tobacco mosaic virus. No abnormalities and no virus present in cells.

Fig. 131 (*above*). Portion of P-protein body associated with plastids and mitochondrion. P-protein tubules are aligned longitudinally in the cell and alternate with rows of small granules. Cell wall shows a loosely structured layer under plasmalemma. × 45,000.

Fig. 132 (*below*). Partly dispersed P-protein clearly showing granules associated with the tubules. × 45,000.

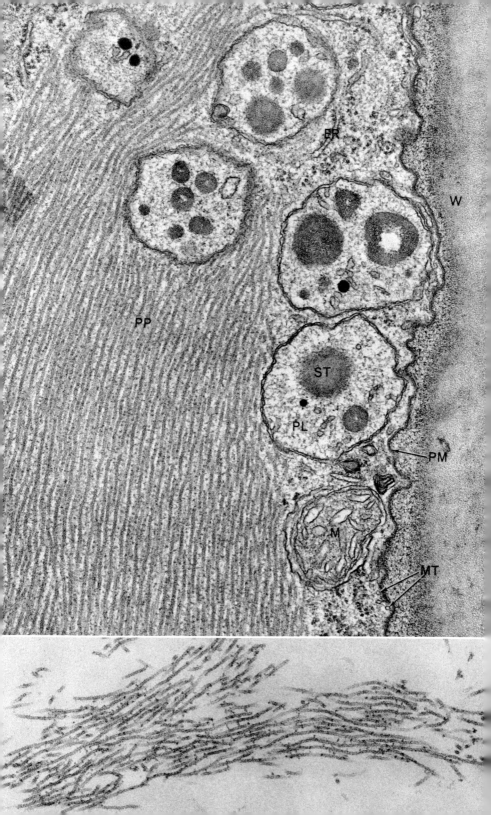

Figs. 133, 134. Longitudinal sections of phloem from small veins of leaves of *Nicotiana tabacum* infected with tobacco mosaic virus. No abnormalities and no virus are discernible in cells.

Fig. 133 (*above*). Parenchyma cell in center is contiguous to sieve element. It contains usual organelles and a small P-protein body. × 7,500.

Fig. 134 (*below*). Like Figure 133, but sieve element contiguous to cell with P-protein body is not visible. × 7,500.

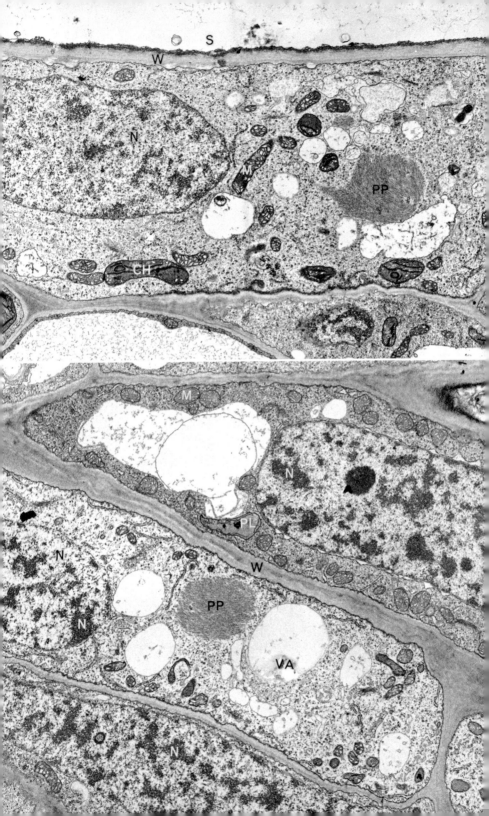

Fig. 135. Longitudinal section of phloem from small vein of leaf of *Nicotiana tabacum* infected with tobacco mosaic virus. No virus is discernible in cells. Parenchyma cell in center is contiguous to sieve element. It contains an unusually large aggregate of P-protein with most of the tubules longitudinally oriented in the cell. They are curved near the vacuole. × 14,000.

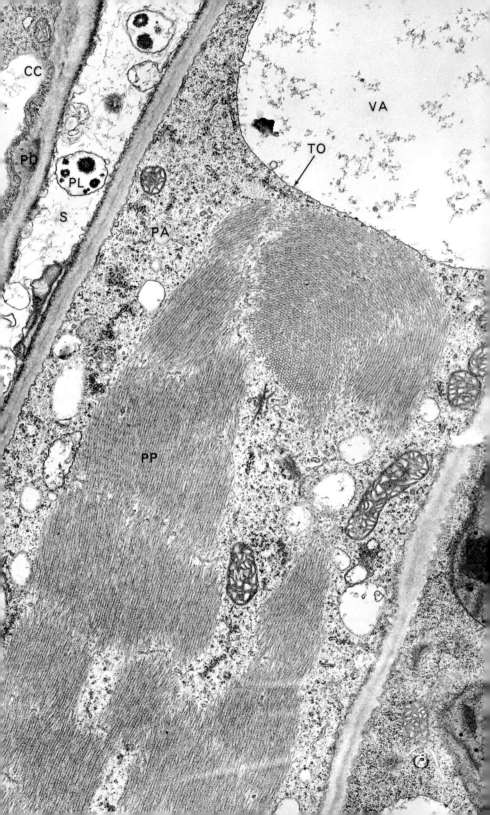

Figs. 136, 137. P-protein from phloem-parenchyma cells similar to the one shown in Figure 135. From small vein in leaf of *Nicotiana tabacum* infected with tobacco mosaic virus. No virus is discernible in cells.

Fig. 136 (*large picture*). P-protein tubules are variously oriented. Where they were cut longitudinally, rows of granules alternating with tubules were exposed. × 53,000.

Fig. 137 (*insert*). P-protein tubules from aggregate similar to the one in Figure 136. Shown in transection at same magnification as P-protein tubules in Figure 72 and X-tubules in Figure 20. × 120,000.

INDEX

Ameboid body. *See* X-body

Anisometric virus: definition of, 18

Arabis mosaic virus: effect on endoplasmic reticulum, 68

Artichoke mottled crinkle virus: in cytoplasm, 19, 23, 24; in vacuole, 23; in tracheary elements, 37

Barley stripe mosaic virus: particle structure, 11; in nucleus, 25; effect on chloroplasts, 51

Beet yellows virus: relation to host tissues, 4, 32; source of infected material, 5; intracellular form, 17–18, 23, 24, 25, 112–117, 122, 123, 150–151; in parenchyma cell, 17, 23, 24, 25, 112–117, 122, 123, 150, 151; in cytoplasm, 17, 24, 25, 112–117, 122, 123, 150, 151; particle structure, 18; in vacuoles, 23; in cytoplasmic protrusions, 24, 25, 66, 67, 122, 123, 182, 183; in nucleus, 25, 67, 116, 117, 198, 199; relation to cell type, 36; in sieve element, 36, 41, 42, 43, 44, 66, 67, 150–151, 158, 159, 160–161, 198–199; in sieve-plate pores, 41, 42, 43, 158, 159, 160–161; in plasmodesmata, 41, 42, 158–159; effect on chloroplasts, 52–55, 58, 60, 174–179; effect on sieve-element plastids, 62, 63, 64, 190–191; effect on phloem, 66–70, 196–205

Beet yellows virus infection: accumulation of osmiophilic globules in, 54–55, 57, 176–183; accumulation of phytoferritin in, 58, 59, 182, 183; accumulation of starch in, 67, 68, 178, 179, 182, 183; inducing degeneration into vesicles, 68, 69, 198–205 *passim*; inducing degeneration into flaky material, 69, 176, 177, 200, 201, 202, 203. *See also* Beet yellows virus

221

Brazilian tomato spotted wilt virus: particles in relation to plasmodesmata, 42

Broadbean mottle virus: crystallization, 18

Callose, 41, 152–153, 155–157, 159–161 *passim*

Chloroplast: affected by tobacco mosaic virus, 27, 28, 51, 52, 53, 60–62, 130–131, 184–187; containing tobacco mosaic virus particles, 27, 28, 61, 130, 131, 186, 187; affected by viruses, 51–62, 130–131, 174–179, 184–187; affected by beet yellows virus, 52–55, 58, 60, 174–179; structure in *Beta vulgaris*, 52, 172–173, 174, 175; structure in *Nicotiana tabacum*, 61, 184, 185

Citrus tristeza virus: size of particle, 18; particles in phloem, 36

Conducting cells: relation to virus transport, 38. *See also* Sieve element; Tracheary element

Cowpea chlorotic mottle virus: protein of, 15; hybridization with tobacco mosaic virus, 16

Crystalline tobacco mosaic virus: growth, 8; arrangement of particles in, 8, 9, 92–95; monolayer in, 8, 92, 93; herringbone pattern, 9, 92, 93; needle-shaped, 10, 96, 97; in intercellular space, 25, 122, 123; penetration of nucleus, 27

Curly top virus. *See* Sugar beet curly top virus

Cytokinesis: in cell infected with tobacco mosaic virus, 45–47, 162–163, 166–171

Enation mosaic virus. *See* Pea enation mosaic virus

Flaky material: in beet yellows virus infection, 69, 176, 177, 200, 201, 202, 203

Isometric virus: definition of, 18

Kinetochore: mentioned, 126, 127

Lettuce necrotic yellows virus: in tracheary element, 37

Lipid globules. *See* Osmiophilic globules

Lomasome: mentioned, 118, 119

Maize rough dwarf virus: in endoplasmic reticulum, 28; viroplasm in infected cell, 29; in sieve element, 37; effect on chloroplast, 52

Material used for study, 4–6

Methods used in study, 5

Microbody: mentioned, 172, 173, 186, 187

Microtubules: compared with X-material, 15, 110, 111; in phragmoplast, 47, 170–171; compared with P-protein, 66

Mitosis: in cell infected with tobacco mosaic virus, 27, 126–129, 164–165

Monolayer: in crystal of tobacco mosaic virus, 8, 92, 93

Mosaic virus (unnamed) in *Lantana horrida*: particle structure, 12; in nucleus, 25

Mosaic virus (unnamed) in sunflower: effect on chloroplast, 51, 53, 61

Mottled crinkle virus. *See* Artichoke mottled crinkle virus

Necrotic yellows virus. *See* Lettuce necrotic yellows virus
Nuclear pores: structure, 26, 124, 125

Osmiophilic globules: in chloroplasts, 53–58, 61, 176–183; in cytoplasm, 54, 57, 180, 181, 182, 183; in sieve-element plastids, 63, 190, 191

Pea enation mosaic virus: in plant and insect vector, 19; in nucleus, 25
Petunia ringspot virus: effect on dictyosomes, 68
Phloem: affected by viruses, 64–70, 150–151, 196–205, 208–211, 216–219; structure in *Beta vulgaris*, 65, 66, 192–195; affected by beet yellows virus, 66–70, 150–151, 196–205
Phragmosome, 164, 165
Phytoferritin: in noninfected plants 58, 59, 60; in *Beta vulgaris* infected with beet yellows virus, 58, 59, 182, 183
Plasmodesmata: in guard cells, 33, 134, 135; structure, 39–40, 152–155; as paths of virus transport, 41, 42, 158, 159
Plastids: containing particles of tobacco mosaic virus, 27, 28, 34, 61, 64, 130, 131, 132, 133, 186, 187; affected by tobacco mosaic virus, 27–28, 34, 51, 52, 53, 60–62, 64, 130–133, 184–187; affected by beet yellows virus, 52–55, 58, 60, 62, 63, 64, 174–179, 190–191; affected by curly top virus, 62. *See also* Chloroplast; Sieve-element plastid
P-protein: in young sieve element, 34, 35, 66, 72, 132, 133, 142, 143, 144, 145, 208–209; relation to sieve plate, 41, 42, 43, 44, 156–157, 158, 159, 160–161; associated with beet yellows virus particles, 42, 43, 158, 159, 160–161; on sieve-element plastid, 63, 64, 188, 189, 190, 191; compared with microtubules, 66; compared with X-material, 70–73, 210–211; structure, 71–73, 156–157, 210–213, 218–219; in parenchyma cell, 72, 73, 210–211, 214–219
Protrusions of cytoplasm into vacuole: containing various viruses, 23–25; containing beet yellows virus, 24, 25, 66, 67, 122, 123, 182, 183; containing tobacco mosaic virus, 24, 118–123

Rattle virus. *See* Tobacco rattle virus
Rice dwarf virus: in host plant and vector, 19, 20
Ringspot virus. *See Petunia* ringspot virus
Rough dwarf virus. *See* Maize rough dwarf virus

Sieve area. *See* Sieve plate
Sieve element: containing tobacco mosaic virus, 33–36, 44, 138–145; containing beet yellows virus, 36, 41, 42, 43, 44, 66, 67, 150–151, 158, 159, 160–161, 198–199; in *Beta vulgaris* leaf, 194–195; in *Nicotiana tabacum* leaf, 208–209. *See also* Phloem; P-protein; Sieve-element plastid
Sieve-element plastid: structure in *Nicotiana tabacum*, 28, 132,

133, 212, 213; affected by tobacco mosaic virus, 34, 64, 132, 133; affected by curly top virus, 62; affected by beet yellows virus, 62, 63, 64, 190–191; structure in *Beta vulgaris*, 62–63, 188–189

Sieve plate: structure, 40–41, 156–157, 158, 159; with beet yellows virus in pores, 41, 42, 43, 158, 159, 160–161

Starch in relation to viral infections, 51, 52, 58, 67, 68, 178, 179 *passim*

Striate mosaic virus. See Wheat striate mosaic virus

Stripe mosaic virus. See Barley stripe mosaic virus

Sugar beet curly top virus: effect on sieve-element plastid, 62

Tobacco etch virus: inclusions of, 18; in mixed infection with tobacco mosaic virus, 19

Tobacco mosaic virus: relation to host tissues, 4, 32; source of infected material, 5; intracellular form in cytoplasm, 7–17, 19, 23–24, 92–97, 118–123, 162–163, 206–207; intracellular form in parenchyma cell, 7–17, 19, 23–24, 92–97, 118–123, 162–163, 206–207; types of inclusions in host cells, 8; linear aggregation of particles, 9–10, 94–97, 140–141; particle structure, 11–12, 98–101; in X-body, 14, 102–103, 108–109; hybridization with cowpea chlorotic mottle virus, 16; in mixed infection with tobacco etch virus, 19; multiplication in host cell, 21, 22, 26; in vacuole, 23–24, 118–119, 208–209; in cytoplasmic protrusions, 24, 118–123; in intercellular space, 25, 122, 123; relation to nucleus, 25–27, 36, 124–129, 148, 149; effect on chloroplast, 27, 28, 51, 52, 53, 60–62, 130–131, 184–187; in chloroplasts, 27, 28, 61, 130, 131, 186, 187; behavior in mitosis, 27, 126–129, 164–165; in guard cell, 32–33, 134–135; in sieve element, 33–36, 44, 138–145; in conducting cells, 33–37; in crystalliferous cell, 33, 136–137; effect on sieve-element plastids, 34, 64, 132, 133; in tracheary element, 35, 36, 37, 146–149; relation to pores in walls, 42; behavior in cytokinesis, 45–47, 162–163, 166–171; effect on mesophyll, 60, 61

Tobacco mosaic virus crystal. See Crystalline tobacco mosaic virus

Tobacco rattle virus: effect on chloroplasts, 51

Tomato spotted wilt virus. See Brazilian tomato spotted wilt virus

Tracheary element: containing tobacco mosaic virus, 35, 36, 37, 146–149; disintegration of primary wall in, 35, 146, 147

Tristeza virus. See Citrus tristeza virus

Turnip yellow mosaic virus: protein of, 15; in sieve element, 37; effect on chloroplasts, 51

Vacuolate body. See X-body

Vesicles in beet yellows virus infection, 68, 69, 198–205

Viroplasm, 29–30
Virus (unnamed) in *Amaranthus lividus*: effect on chloroplasts, 53
Viruses: first identifications with electron microscope, 7; intracellular form, 7–20; anisometric, 18; isometric, 18; comparison of in plant and insect vector, 19–20; inclusions of in host cells, 20; sites of synthesis in host cell, 21, 22, 26, 29; relation to host-cell components, 21–30; occurrence in vacuoles, 23, 24; in cytoplasmic protrusions, 23–25; crystallization in vitro, 24; relation to host tissues, 31–32; relation to cell types, 31–37; intercellular transport, 38–48; movement in phloem, 43–44; movement in xylem, 44, 45; nature of infection, 49–50, 71; pathologic effects, 49–73; effect on chloroplasts, 51–62; effect on phloem, 64–70
Virus-related material, 12, 13, 15, 16. *See also* X-body; X-material; X-protein

Watermelon mosaic virus: linear aggregation of particles, 11; inclusions, 18
Wheat striate mosaic virus: in nucleus, 25; particles in relation to plasmodesmata, 42

Wound tumor virus: comparison in plant and insect vector, 19; in endoplasmic reticulum, 28; viroplasm, 29–30; particles in various cell types, 37

X-body: interpretation of, 8, 12, 13; structure, 13–15, 102–103, 108–109, 206, 207; development, 13–15, 104–105, 108–109; in nucleus, 25. *See also* X-material
X-component. *See* X-material
X-material: interpretation of, 8, 12–17; development, 13–15, 104–111; compared with microtubules, 15, 110, 111; compared with tobacco mosaic virus particle, 16, 106, 107; in cytoplasm, 23, 102–111; in parenchyma cell, 23, 102–111, 206, 207, 210–211; in nucleus, 25; relation to cell type, 33; in sieve element, 35, 144–145; in tracheary element, 36, 148, 149; in cytokinesis, 47, 170–171; compared with P-protein, 70–73, 210–211. *See also* X-body
X-protein, 12, 13, 15
X-tubules. *See* X-body; X-material

Yellow mosaic virus. *See* Turnip yellow mosaic virus